THE NEUROBIOLOGY OF CRIMINAL BEHAVIOR

THE NEUROBIOLOGY OF CRIMINAL BEHAVIOR

edited by

Joseph Glicksohn
Bar-Ilan University, Israel

KLUWER ACADEMIC PUBLISHERS
Boston / Dordrecht / London

Distributors for North, Central and South America:
Kluwer Academic Publishers
101 Philip Drive
Assinippi Park
Norwell, Massachusetts 02061 USA
Telephone (781) 871-6600
Fax (781) 681-9045
E-Mail: kluwer@wkap.com

Distributors for all other countries:
Kluwer Academic Publishers Group
Post Office Box 322
3300 AH Dordrecht, THE NETHERLANDS
Telephone 31 786 576 000
Fax 31 786 576 474
E-Mail: services@wkap.nl

Electronic Services <http://www.wkap.nl>

Library of Congress Cataloging-in-Publication Data

0-7923-7674-9
A C.I.P. Catalogue record for this book is available
from the Library of Congress.

Printed on acid-free paper.

Printed in the United States of America.

To my loving family: Chanita, Arit and Ofer

TABLE OF CONTENTS

Part D Neurocognitive Models and Psychobiological Markers for Criminality

Acknowledgments

An edited book is only as good as the caliber of its authors. In the present case, it was gratifying to work with such a fine team of contributors. It is also a pleasure to thank Michael Myslobodsky for encouraging me to contribute an edited volume for his series on the neurobiology of aberrant behaviour. We go back a long way: It was an honour to do a postdoctorate with him less than a decade ago. Since then, I have been thinking about the neurobiology of criminal behaviour, trying to weave in a number of lines of research. My own biases will be laid bare in my introductory chapter.

Three contributors (Marvin Zuckerman, James Blair and Edgar Miller) presented their material at a symposium on the neurobiology and neuropsychology of criminal behaviour, which I co-chaired with Vincent Egan, at the recent meeting of the International Society for the Study of Individual Differences, in Edinburgh (July, 2001). My thanks to Ian Deary for his encouragement and support of that quite successful session.

In editing this volume, I suggested to the contributors that we conduct an "in house" and open review, whereby the various authors would read each other's manuscripts. This suggestion was enthusiastically taken up by all concerned, and was implemented as far as possible, given the time constraints and heavy schedules of everyone. In addition to the authors themselves, I would like to thank Ernest Barratt and Gerard Moeller for helping out with this review process. My own chapter was reviewed by Michael Myslobodsky and Israel Nachson, both of whom provided me with insightful comments. Chanita Goodblatt was both a source of support and of informed criticism. My thanks to Michael Williams and Mary Panarelli at Kluwer Academic Publishers for their helpful suggestions.

Pamela Taylor has provided an exquisite Foreword. Based on what she writes, this is a volume well worth reading. Need I say more?

LIST OF CONTRIBUTORS

Alan Apter
Department of Psychiatry
Tel-Aviv University
Ramat-Aviv 69972, Israel

Robert H. Belmaker
Ben Gurion University of the Negev
PO Box 4600
Beersheva, Israel

R. James R. Blair
Institute of Cognitive Neuroscience
University College London
Alexandra House
17 Queen Square
London WC1N 3AR, England

Richard P. Ebstein
Herzog Hospital
PO Box 35300
Jerusalem 91351, Israel

Adrienne K. Elliott
Section on Clinical and
Experimental Neuropsychology
NIMH/NIH
5415 W. Cedar Lane
Suite 203B
Bethesda, MD, USA

Joseph Glicksohn
Department of Criminology
Bar-Ilan University
Ramat-Gan 52900, Israel

Sharon S. Ishikawa
Department of Psychology
S.G.M. Building
University of Southern California
Los Angeles, CA 90089-1061, USA

Edgar Miller
Center for Applied Psychology
University of Leicester
Leicester LE1 7RH, England

Allan F. Mirsky
Section on Clinical and
Experimental Neuropsychology
NIMH/NIH
5415 W. Cedar Lane
Suite 203B
Bethesda, MD, USA

Adrian Raine
Department of Psychology
S.G.M. Building
University of Southern California
Los Angeles, CA 90089-1061, USA

Gil Zalsman
Adolescent Inpatient Unit
Geha Mental Health Center
P.O. Box 102, Petach Tikva 49100,
Israel

Marvin Zuckerman
Department of Psychology
University of Delaware
Newark, DE 19716-2577, USA

Foreword

THE NEUROBIOLOGY OF CRIMINAL BEHAVIOUR

Pamela J. Taylor, MD
Department of Forensic Psychiatry, Institute of Psychiatry, London

INTRODUCTION

Legal systems vary from nation to nation, and over time. In general, though, for an act to be regarded as a crime, it is necessary to prove that the accused carried out the act, and that he was in a state of mind to have the capacity to commit the crime in question. Capacity explicitly incorporates cognitive qualities—the ability to know what one is doing and/or whether it is wrong—and volitional qualities—the ability to control physical acts in accordance with rational judgment. Any impairment in these areas must, at some level, be attributable to neurobiology. Given potential status in law, however, it is apparent that neurobiological explanations within this field go beyond conventional clinical boundaries.

Allowance in criminal law for neurobiological status may be at various levels. Developmental stage, for example, is almost invariably taken into account. In most jurisdictions, individuals under a legally designated age cannot be tried for or convicted of a crime; a few societies set an upper age limit too. Most allow that in the mature person, disease may alter capacity in this respect, and some allow for degrees of its impairment. Submitting neurobiological evidence in Court is, however, rarely simple, even in the presence of undisputed evidence of brain damage (e.g., Taylor, 1985). Denno (1996) reviewed some of the arguments in relation more specifically to genetic evidence. It is inevitable and proper that clinicians and scientists seek to define not only relevant presentations in mental state—impairments of mind—but also, where relevant, their causative physical conditions; it is also reassuring that, as yet, Courts retain doubt about inferring causal connections from scientific evidence of possible associations.

This book constitutes another important step in furthering an understanding of the nature and quality of association between neurobiological status and criminal acts. Its strengths are considerable, as will be readily apparent from the calibre of the contributors and their thoughtful presentation of data in each chapter. Many of the authors are explicitly cautious about the case, as yet, for translation of findings—their own and those of others—into practice. As the risks of premature use of such material and its politicization are greater in this field than almost any other in the biological sciences, such cautions would bear repetition and I shall dwell disproportionately on the limitations of research in this field.

CRIMINAL BEHAVIOUR AS A POOR CRITERION VARIABLE

The validity of measurement of any behavioural construct poses perhaps the greatest limitation on research in this area. Intelligence, for example, as measured by IQ scales, is generally considered to be one of the more robust measures of an aspect of behaviour, and yet it is not culture free. Herrnstein and Murray's (1994) book, *The Bell Shaped Curve*, demonstrates some of the problems in determining intrinsic biological contributions to intelligence in isolation from environment, and then inferring causation from biological predisposition rather than socially-limited opportunity. In *The Mismeasure of Man*, Steven Jay Gould (1981) had already argued that such measures were designed for political purposes. Lane (1994) demonstrated perhaps less obvious, but equally worrying, politics in the funding and support of such research.

The relationship between politics and science is an area of which everyone working scientifically must be aware. There are currently some particularly relevant issues in the UK. Academic units are judged by their ability to attract external funds for research; the government in England has decided that there is an as yet undefined group of people—"dangerous with severe personality disorder (DSPD)". It has set aside substantial sums of money for research into the assessment and management of this hypothetical group. All intentions seem honorable in the application for and allocation of monies, but it is not difficult to see how these could slip. The concept of DSPD is accompanied by proposals for a whole new set of services and legal provisions to lock up and control people who may fall into the category.

Criminal behaviour is a social concept, and, as such, subject to the vicissitudes of society and time. Crimes in some societies are not crimes in others, and political dissent is easily criminalised. Then, within any one society, an act may be a crime on one day, and not on the next. In the UK, for example, on 2nd August 1961, suicide was a crime; on 3rd August 1961, it was not. On 16th July 1967, homosexual acts were unlawful, by 17th July 1967, it was lawful for men of 21 and over to have homosexual relations in private. Such forms of behaviour, and those which have been perhaps more consistently designated as anti-social or criminal, are almost universal. So, in what are we really interested? Is there a common core of behaviour that can universally and reliably be designated as anti-social or criminal? Even if so, what possible rationale could there be for a hypothesis that any category of crime would be associated with a unique neurobiological state?

Supposing that there could be agreement that homicide, child sexual assault and theft constitute criminal behaviours, there is a further problem. Each is very different from the others and none constitutes a homogenous type of behaviour. As complex human behaviours, none could. Scientific solutions to this attempt to deconstruct the behaviours into presumed component parts—but this creates new problems, since it is arguable that such laboratory models bear little relationship to reality. We even fall into a state of circularity, setting tasks to improve primary definitions of the phenotype, while supposing also that biological markers can help us with those definitions.

The use of crime as a target variable does have an important theoretical strength; evidence that the behaviour actually occurred has generally been thoroughly tested. Errors and miscarriages of justice are reported, however, and only a small proportion of reported crime is ever cleared up. Then, in spite of all the factors thus mitigating against it, in many communities it is far from unusual for people to have a criminal record. In England and Wales, for example, 40% of men will have been convicted of a criminal offence by the age of 50 (Farrington, 1981), more in the USA (Blumstein & Cohen, 1987). Criminal behaviour per se can thus hardly be regarded as abnormal. This has not stopped some researchers from using the broad concept as a criterion, but has led others to target crime for these purposes in terms of seriousness, frequency or both. The latter, pragmatic approach, has face validity, but is that enough?

In summary, there seems something intrinsically mistaken about any hypothesis which posits any general relationship between criminal behaviour and neurobiology. It seems plausible to suppose that neurobiology could and should be invoked in more specific hypotheses, perhaps particularly in individual cases. Neurobiological mediation in cases, for example, when a crime occurred in the context of mental disorder, but not otherwise, would be a prime target for hypothesis formation. Indeed, at least since the English landmark case of Hadfield in 1800, we have had just such an incentive. He attempted to assassinate the then king, George III, an act which led to the Act for the Safe Custody of Insane Persons charged with Offences 1800, the precursor of modern English legislation in this area. The critical point here is that there is no suggestion that assassination in general implies disorder, nor that disorder of a particular kind will cause a person to become an assassin, but rather the possibility of causative pathways in individuals. Hadfield had been a skilled craftsman and model soldier until he sustained penetrating head injuries in battle. Thereafter, his characteristic behaviour changed. It is something of a puzzle to me that the weight of research testing associations between criminal behaviour and neurobiological deficits is with groups of people who fall towards the end of a dimension or a normal curve. This book includes a chapter on head injury, but there is little reference to people who have developed a disease which becomes a major factor in emergent criminal behaviour. Schizophrenia, for example, has a recognised significant, if small, association with violence, and carries its own neurobiological literature, and yet it does not have a chapter here.

NEUROBIOLOGY LIKELY TO ASSIST IN ONLY A MINORITY OF CASES

Longitudinal cohort studies are and have been invaluable in developing an understanding of pathways into crime. Farrington (1993) reviewed the sixteen studies worldwide which are robust according to a set of criteria he generated. These were: having a sample size of several hundred drawn from the general population, a large number of variables, prospective study over at least five years, and incorporation of self-report, preferably other direct observer measures, as well as official measures such as criminal record. Perhaps the most pertinent of the cohort work here is that of Moffitt, Caspi and others, based principally on the Dunedin birth cohort.

In the Dunedin birth cohort, measures of antisocial behaviour were made every other year between the ages of 3 and 15, again at 18, and at 21, and the study continues. The ratings employed were psychometric, or were made by teachers, parents, and, from age 11, the subjects themselves, together with official criminal and police records. At ages 18 and 21, peer, and where relevant, partner reports were added (Moffitt *et al.*, 2001). It was evident that some subjects showed early childhood onset of antisocial behaviour, but also that by age 13 and 15 new cases of individuals with previously unremarkable development were emerging. Further follow through confirmed two principal patterns of antisocial behaviour: *early-onset, life-course persistent* and *adolescent-onset, time-limited* (Caspi & Moffitt, 1995; Moffitt, 1993, 1997).

Moffitt and Caspi's more recent work has focused on male-female differences, a male preponderance popularly assumed to be one of the most robust findings in criminological research (Moffitt *et al.*, 2001). They found little difference between men and women in the fact, cause, correlates and consequences of crime for the adolescent-limited group. Sociological explanations dominated. The sex ratio for childhood-onset, life-course-persistent antisocial behaviour was, however, 10 males to one female; further, this pattern was associated with neuropsychological differences and indicators of biological maturational delay. This is the group, they suggest, on which to focus research on neuro-biological explanatory models. David Taylor's (1969) recognition of a period of vulnerability to insults such as anoxia and infection on the maturing infant brain may provide a partial explanation of the emergence of such a group. The infant male brain matures more slowly than the infant female brain, and the longer period of vulnerability for the former thus exposes more boys than girls to risk of damage. This would broadly fit with the documented sex ratio for early onset antisocial behaviour, but so it would with the many other developmental disorders which are more common in boys, including autism, Asperger's syndrome and specific language disorders. Neurobiological damage may create a substrate for anti-social behaviour, but it is likely always to be important to allow for other environmental factors which may maintain, shape, or limit its manifestation.

EVIDENCE BASED, CULTURE-FREE CRITERION VARIABLES?

Many of the contributors to this book have adopted measures of personality which incorporate some of these concepts, which have been designed to be relatively culture-free and to avoid the problems of circularity and cultural contamination inherent in defining antisocial personality disorders in terms only of repeated antisocial behaviours. The concept of *psychopathy* (Hare, 1991) is principal among these. While the checklist which underpins this concept generates two dimensions—one essentially reflecting repeated antisocial acts, but the other achieving some independence of these—the extent to which this concept is culture-free remains unclear. Although there is an extensive literature on the reliability and validity of the tool—the *Psychopathy Checklist-Revised* (PCL-R)—it is mainly from North America, with a little from Northern Europe. Insofar as the scale is used categorically, a lower cut-off is recommended in the UK than in North America. How should other countries and cultures interpret work using this measure? Others measures have perhaps been subjected to even less testing. How well do concepts of *sensation seeking* and even *hyperactivity* travel? Undoubtedly, important advances have been made in defining and measuring behaviours which may act as intermediaries between neurological state and crime. Nevertheless, the reader must maintain awareness of questions which remain around such concepts in a critical reading of the collection of work reported here.

ETHICAL ISSUES IN APPLICATIONS OF WORK IN THIS FIELD

Criminal behaviour poses heavy burdens on its victims, on society more generally and, among those who are caught in their criminal behaviour, on the perpetrators too. Research is likely to have a vital role in easing those burdens. While much of that work must rest with criminologists and sociologists, neurobiological research must be a part of a complete picture. Some, including aspects of my past work, may provide little more than indirect corroboration of social disadvantage. What is the meaning or use of a finding, for example, that among convicted male prisoners, men with violence convictions as a group, scored significantly lower on intelligence tests than those without, all scores nevertheless being within the normal range (Robertson *et al.*, 1987)? This may be an important finding, and some programs within the criminal justice system aim to improve general intellectual functions, but statistical significance is not necessarily the same as practical importance. Nevertheless, the idea that neuropsychological findings could lead to remedial cognitive behavioural programs, neurophysiological or neurochemical findings to pharmacological treatments, or a genetic breakthrough to some other finely-directed intervention is inescapably attractive, if unlikely at least in the latter case.

As a clinician, my main interest lies in how much this book has captured for me to offer to individual patients. The answer is not much yet, but extrapolation from group findings to the individual case is rarely straightforward. Further, this field has tended to rely especially on very small descriptive samples; larger, more apparently methodologically sound studies tend to have substantial sample sizes only because effect sizes are so small. Blindness between raters of clinical/behavioural status and raters of neurobiological measures is very important as many are soft, or have the potential for contamination by subjective judgement, but still only a limited number of studies achieve such standards. Still, this book is an encouraging record of how much methodology is improving.

A greater concern is the extent to which the field may be open to misuse. Concepts of both genetic/organic determinism and of social determinism are equally unacceptable in the risks attached of isolation, disadvantageous discrimination and stigmatisation. The research is not, and cannot be value free. In the UK, the Nuffield Council on Bioethics (1998) has already published a relevant report on mental disorder, focussing on genetics research, although many of the same issues would apply to the wider neurobiological field. Publication of a report on the ethical context of such research with respect to human behaviour is pending.

In general, sound and improving systems are in place for evaluating the ethics of research with human subjects. Research proposals must be submitted to an ethics committee of appropriate constitution. In the UK this is now taken to mean a body which includes lay representatives as well as people independent of the proposal but with expertise in scientific methods, and perhaps in the specific areas to be researched. Properly informed lay views are especially important in a field where research findings could be used to deprive people of their liberty, perhaps people who have never committed a criminal act but who, by virtue of a neurobiological marker, are construed as being at risk for doing so. The onus is on the scientific community to explain proposed research, and its potential applications, in language that is simple and clear enough for non-experts to understand. This will be important too so that potential subjects can make adequately informed consent.

There are few comparable systems for evaluating the ethics of application of research findings in practice, although there are precedents. Work on human fertility is one such in some countries. While it is arguable that the freedom to conduct and report ethical research should not be constrained by the possibility that findings could be misused, it does seem that in this field there may be a case for special scrutiny of its applications. The evidence-based concepts of *life-course-persistence* in the context of antisocial behaviour could lead to early intervention programs which could benefit the individual and society alike; a more pessimistic or parsimonious reading of the data could lead to deliberate exclusion from society, or indefinite custodial detention. Apparent further validation by identification of physical correlates or causes could exacerbate this tendency. The effects of presumed genetic pre-disposition already have an alarming history. Then, too, genetic information affects not only the individual concerned, but his or her family as well.

THE IMPORTANCE OF BOOKS, THE IMPORTANCE OF THIS BOOK

Books allow for the more leisurely and thorough contemplation of subjects in the round than any single scientific paper can achieve. Current trends in academia are towards valuing the peer-reviewed, data-based paper above all else. This book embraces such work, but provides a readable synthesis of it. It is sensitive on many of the issues that I have raised. It provides us with a vehicle onto which we can project new ideas and create even better syntheses and new hypotheses. It is an honour to introduce it.

REFERENCES

Blumstein, A., & Cohen, J. (1987). Characterising criminal careers. *Science,* **237**, 985-991.

Caspi, A., & Moffitt, T.E. (1995). The continuity of maladaptive behaviour: From description to explanation of the study of antisocial behaviour. In D.

Cicchetti and D. Cohen (Eds.), *Developmental psychology* Vol. II. New York: Wiley.

Denno, D.W. (1996). Legal implications of genetics and crime research. In M. Rutter (Ed.), *Genetics of criminal and antisocial behaviour* (CIBA Foundation Symposium 194). Wiley: Chichester.

Farrington, D.P. (1981). The prevalence of convictions. *British Journal of Criminology,* **21**, 173-175.

Farrington, D.P. (1993). The psychosocial milieu of the offender. In J. Gunn & P.J. Taylor (Eds.), *Forensic psychiatry: Legal, clinical and ethical issues* (pp. 252-285). Oxford: Butterworth-Heinemann.

Gould, S.J. (1981). *The mismeasure of man.* New York: Norton and Co.

Hare, R.D. (1991). *The Hare psychopathy checklist-revised.* Toronto, Ontario: Multi-Health Systems.

Herrnstein, R.J., & Murray, C. (1994). *The Bell Curve: Intelligence and class structure in American life.* New York: Free Press.

Lane, C. (1994). The tainted sources of *The Bell Curve. New York Review of Books,* **XLI**, 14-19.

Moffitt, T.E. (1993). "Life-course-persistent" and "adolescent-limited" antisocial behaviour: A developmental taxonomy. *Psychological Review*, **100**, 674-701.

Moffitt, T.E. (1997) Adolescence-limited and life course persistent offending: A complementary par of developmental theories. In T. Thornberry (Ed.), *Advances in criminological theory: Developmental theories and crime and delinquency* (pp. 11- 54). London: Transaction Press.

Moffitt, T.E., Caspi, A., Rutter, M., & Silva, P.A. (2001). *Sex differences in antisocial behaviour.* Cambridge: Cambridge University Press.

Nuffield Council on Bioethics (1998). *Mental disorders and genetics: the ethical context.* London: Nuffield Council.

Nuffield Council on Bioethics (In press). *Genetics and human behaviour: the ethical context.* London: Nuffield Council.

Robertson, G., Taylor, P.J., & Gunn J. (1987). Does violence have cognitive correlates? *British Journal of Psychiatry*, **151**, 674-678.

Taylor, D.C. (1969). Differential rates of maturation between sexes and hemispheres. *Lancet*, **ii**, 140-142.

Taylor, P.J. (1985). Epilepsy and insanity. In P. Fenwick & E. Fenwick (Eds.), *Epilepsy and the law.* London: Royal Society of Medicine.

PART A

INTRODUCTION

Chapter 1

CRIMINALITY, PERSONALITY AND COGNITIVE NEUROSCIENCE

Joseph Glicksohn, Ph.D.
Department of Criminology, Bar-Ilan University

INTRODUCTION

On entering the 21st century, the "decade of the brain" (see http://www.loc.gov/loc/brain) has now been followed by a "decade of behaviour" (see http://www.decadeofbehavior.org). It is thus of topical interest to consider the *neurobiology* of criminal *behaviour*, an endeavour spanning these two foci of research interest. Criminal behaviour will be looked at in the widest sense of the term, and with reference to a variety of neurobiological methodologies of contemporary interest. Our intended audience is drawn from two groups: (1) researchers in the fields of neurobiology, psychology and cognitive neuroscience, who would be interested in learning about current research in criminology and biological psychiatry; and (2) researchers in the latter fields, who would be interested in updating their knowledge to incorporate a neurobiological approach to the study of criminal behaviour.

This volume has 4 parts, with the present chapter serving as an Introduction (Part A). In later sections of the chapter, I consider some facets of intellectual history, together with developments paving the way for a psychobiological approach to criminality and for a neurocognitive research strategy. Part B of the volume, comprising 2 chapters, is concerned with personality and criminality, emphasizing this psychobiological foundation for criminology. In Chapter 2,

Marvin Zuckerman provides a panoramic overview of the construct of sensation seeking and its intricate relationship with the construct of psychopathy. His discussion weaves in psychometric, genetic, neuropsychological, and psychopharmacological literatures. In Chapter 3, Richard Ebstein and Robert Belmaker focus on the genetic basis for this sensation seeking and novelty seeking trait, discussing their work describing the dopamine 4 receptor (DRD4) gene.

Part C of the volume comprises three chapters concerned with genetic, environmental and cognitive factors predisposing to crime. In Chapter 4, Sharon Ishikawa and Adrian Raine provide an integrative review of behaviour-genetic studies, being careful to stress that genetic research can well help to identify the biological mechanisms that shape antisocial behaviour (e.g., poor functioning of the prefrontal cortex, hypoarousal). In Chapter 5, Adrienne Elliott and Allan Mirsky consider cognitive antecedents of violence and aggression which must necessarily moderate such a genetic predisposition for crime. In Chapter 6, Edgar Miller considers brain injury as a contributory factor in offending.

Part D comprises three chapters which provide integrative reviews of the neurocognitive models and psychobiological markers for criminality. In Chapter 7, James Blair contrasts acquired sociopathy and developmental psychopathy. He discusses two recent neurocognitive models, the Somatic Marker model and the Social Response Reversal one, both of which guide neurobiological investigations of crime and its predisposing traits. In Chapter 8, Sharon Ishikawa and Adrian Raine provide an extensive coverage of the psychophysiological literature. Their take-home message is one of acute hypoarousal, be this cortical underarousal (especially in the frontal cortex, together with fronto-temporal abnormalities in the EEG) and electrodermal underarousal. In Chapter 9, Gil Zalsman and Alan Apter discuss the relationship between serotonergic metabolism and violence/aggression. Their take-home message is one of an altered (lowered) serotonin metabolism predisposing to violence.

With this outline of the book in mind, let me stress what the book does not intend to accomplish: A reductionist analysis ("everything is in the genes"). Zuckerman (1992) discusses the various levels implied by a psychobiological approach to personality: traits, social behaviour, conditioning, physiology, biochemistry, neurology and genetics, in this order. The same levels are equally applicable to a neurobiological approach to criminal behaviour, though I would add a cognitive level high on the list. As Zuckerman (1992, p. 681) is careful to note, this is not a reductionist model, for while each level can be studied in isolation, for a better understanding one cannot ignore the other levels. This has been the implicit agenda for the authors of this volume. Thus, in discussing genetic predisposition, the authors of the various chapters have consistently referred to moderating environmental factors; and in discussing cognitive manifestations of a personality trait, the authors of the various chapters have also addressed neurobiological circuitry. The chapters are cross-referenced, and preliminary discussions which are elaborated on in another chapter are noted throughout.

Lao Tzu (Blakney, 1955, p. 100) offers the following insight:

> The world may be known
> Without leaving the house;
> The Way may be seen
> Apart from the windows.
> The further you go,
> The less you will know.

I would suggest that in order to know where one is going, it is important to reflect whence one has come.

BACK TO THE PAST—EARLY SCIENTIFIC TRENDS IN THE STUDY OF CRIMINALITY

One hundred and fifty years ago, Phineas Gage bequeathed in himself a fascinating portrait for research in both personality and neuropsychology (Damasio, 1994). The tamping rod which passed through the ventromedial region of his frontal lobes left him alive, though drastically changed in personality: He was markedly more extraverted, impulsive and aggressive. Extraversion, impulsivity, frontal-lobe dysfunction and aggression are all hallmarks of criminal behaviour. Criminological theory dating back a hundred years was well aware of the need to develop a neurobiology of such behaviour, invoking notions of degeneration, neuroanatomical aberration and neurocognitive anomalies. Indeed, talk of evolution was in vogue. While the more fortunate members of society made the transition to higher and higher levels of intellectual and social development, the less fortunate ones might exhibit arrested development at best (Fink, 1985, p. 99), and perhaps a reversion to a more primitive mode of being at worst (Fink, 1985, p. 104).

What was (and is) the basis for such a developmental viewpoint? As Glicksohn (1995) has suggested (in another context), it is only within the framework of a nonlinear developmental theory, such as the organismic-developmental theory of Heinz Werner (1948, 1957/1978), that such notions as developmental arrest, regression and reversion have meaning. Recall that Werner (1948) suggested that less developed systems become subordinated to, and regulated by, more developed ones, but are never actually lost. Normally, these former "drafts" (to use the term suggested by Dennett & Kinsbourne, 1992) are not readily accessible, unless the system is faced with a stressful situation, novel or difficult task, or suffers trauma (as in the case of Phineas Gage). In the event that the cognitive system is not able to function at such a (normally) higher level, then all is not lost, and the system reverts to a more primitive mode of functioning, much as one sees in the reversion from an abstract to a concrete mode of thinking in both neurologically-impaired patients and in psychopathology (e.g., Watson, Plemel, & Burke, 1979; Wright, 1975). Akin to this form of theorizing is that of the nineteenth-century neurologist Hughlings Jackson (Taylor, 1958), who suggested that the neurocognitive manifestation of such a reversal would be the disinhibition (i.e., release) of more primitive brain functioning (e.g., that of the limbic system) due to cortical dysfunction. As Capruso and Levin (1995, p. 216) suggest, employing contemporary criteria (i.e., DSM-IV; APA, 1994) in a retrospective manner, "Phineas Gage ... would have been diagnosed with personality change due to head trauma, disinhibited type."

At present, one finds a major influence of Jacksonian and neo-Jacksonian ideas in the study of schizophrenia (for a review, see Andreasen, Arndt, Alliger & Flaum, 1995). As Glicksohn, Alon, Perlmutter and Purisman (2000-2001) have noted, Werner (1948) adopted Jackson's evolutionary levels in his own organismic-developmental theory. Both Jackson (Taylor, 1958) and Werner (1948) have suggested that there are different *levels* of cognitive functioning (Werner, 1957/1978). Given this, it is a short way to arguing that criminal behaviour is indicative of a reversion to a lower level of functioning (Lombroso, 1906/1968), either because of neurological insult, social deprivation, the use of drugs, or other environmental influences hampering normative development.

Cesare Lombroso (1901, 1906/1968), discussing both criminals and geniuses, invokes the same form of developmental thinking, as do Jackson and Werner, being particularly fond of such euphemisms as *atavism* and *degeneration*, between which he distinguished. Even so, as Carlson, (1985, p. 135) writes, "whether the criminal demonstrated atavism, an active reversion to an earlier biological type, or arrested development, which blocked any progress beyond primitive capabilities, the outcome was the same." Indeed,

Fink (1985), in reviewing the neurobiological (or, to use the term then in vogue, "criminal-anthropological") approaches to criminology of the 19th and early 20th centuries, has come to the conclusion that "... these attempted explanations of criminal behavior—atavism, its sub-category arrested development, and degeneration—reveals no clear-cut distinctions between them" (p. 105). Nevertheless, the theoretical distinctions are still worthy of consideration. As Marvin Wolfgang (1972, p. 247) writes, "the concept of atavism ... postulated a reversion to a primitive or subhuman type of man, characterized physically by a variety of inferior morphological features reminiscent of apes and lower primates...". In contrast, degeneration implicated "diseased ancestral elements which ceased to evolve progressively and give evidence of the process of devolution, so that pathological individuals manifest rudimentary physical and mental attributes of primitive man" (Wolfgang, 1972, p. 247). While the concept of atavism has come under attack over the past 100 years, invariably so in the introductory course in criminology and its accompanying textbook, that of degeneration is still alive and kicking, and having a major influence on contemporary research. Let us take a closer look at degeneration.

Firstly, as Fink (1985, p. 103) reminds us, "Inasmuch as tissue building depends upon the functional integrity of the nervous system, it is evident that degradation of development—i.e., degeneracy—has a neuropathic foundation." Thus, a neurobiological approach to criminality can well invoke the concept of degeneration, if this can help to foster fruitful research. Secondly, degeneration implicates disinhibition, as suggested by Jackson, and as more recently iterated by Colin Martindale (1971). But there is more to the concept, as Martindale (1993, p. 209) has argued in discussing Hans Eysenck's model of personality and creativity (e.g., Eysenck, 1993a): "... Eysenck's model is extremely close to the degeneration theories of genius proposed by Lombroso (1895) and Nordau (1895). The gist of these theories is that degeneration (cf. psychoticism) predisposes one to criminality, psychoses of all types, and genius..." Thus, we readily encounter the notion over the past 30 years, albeit translated into such contemporary terms as disinhibition or psychoticism (see also Martindale, 1999). If Martindale is correct in his assessment, then the concept is still a viable one, especially when viewed via a neurocognitive lens. But 19th century discussions of degeneration quickly degenerated into *fin de siècle* hysteria (some of which the present readership can appreciate, given our own mass hysteria regarding the Y2K bug).

Degeneration and degenerates were hot topics for the Parisian and Viennese salons at the turn of the 20th century (Harris, 1989). Max Nordau (1895) devoted a book to the topic, which he dedicated to Lombroso. Here are some characteristic sections:

> When this phenomenon is present in a high degree, we speak of 'moral insanity'.... The two psychological roots of moral insanity, in all its degrees of development, are firstly, unbounded egoism, and, secondly, impulsiveness—i.e., inability to resist a sudden impulse to any deed; and these characteristics also constitute the chief intellectual stigmata of degenerates. (pp. 18-19)
> There is here a disastrous, vicious circle of reciprocal effects. The drinker (and apparently the smoker also) begets enfeebled children, hereditarily fatigued or degenerated, and these drink and smoke in their turn, because they are fatigued. (p. 41)

Clearly, society was at great risk from these degenerate criminals. But mass hysteria does not confine itself to one topic area, rather it progressively spreads from one domain to another (Showalter, 1997). Positive forms of deviant experience (e.g., creativity, art, synaesthesia), currently being studied in connection with psychoticism, were maligned as much as were negative forms of deviation (criminality), because both could be attributed to degeneration (Dann, 1998).

Perhaps the more surprising thing is that instead of discussing this plausible neurobiological approach to criminality, there was a mass stampede out of the arena, in an attempt to do away with biological explanations. Conceivably, Lombroso's concept of atavism was too threatening to a society still reeling from the implications of the Darwinian revolution, and the Man-Ape comparisons still being discussed. As Charles reveals to Ernestina, in John Fowles' novel, *The French Lieutenant's Woman*, set in Victorian England of 1867, he had a "small philosophical disagreement" with her father: "Your father ventured the opinion that Mr Darwin should be exhibited in a cage in the zoological gardens. In the monkey-house. I tied to explain some of the scientific arguments behind the Darwinian position. I was unsuccessful." (Fowles, 1970, p. 11). The concept of degeneration probably became guilty by virtue of association.

PERILS OF INTERDISCIPLINARITY: THE SOCIOLOGICAL BACKLASH AND THE BIOLOGICAL TABOO

Just as twentieth-century psychology, particularly the American brand of Behaviorism, saw no need to resort to either cognitive and/or neurobiological interpretations of manifest behaviour (e.g., Leahey, 1992), so twentieth-century criminology, particularly the American school of criminology and its favorite textbook (Sutherland & Cressey, 1970) essentially forsook this nineteenth-century psychobiological legacy. As Wright and Miller (1998, p. 1) note in the abstract to their review article on this issue, "biological arguments that link biochemistry, genetics and/or neurophysiology to crime have been viewed as taboo: unthinkable and unmentionable." In fact, while this might be viewed by some as the standard sociological backlash on psychological theorizing in criminology (e.g., Jeffery, 1979; Nachshon, 1982), be this in terms of an assault on the vestiges of positivistic thought (Muncie, McLaughlin & Langan, 1996), or on neurobiological reductionism (Wright & Miller, 1998), one might very well argue with the notion that the major authority in American criminology was himself of a sociological persuasion. Edwin Sutherland's influential textbook went through 11 editions, the last one appearing in 1992 (Sutherland, Cressey & Luckenbill, 1992), 42 years after his death! Wright and Miller (1998) have carefully analyzed the impact of this text on criminological thinking. Their reading implicates a standard Behavioristic tract: The emphasis on learning; the disregard for a genetic basis to criminality; the focus on criminal behavior as opposed to a criminal personality. Here are two suppositions posited in the book (Sutherland & Cressey, 1970):

> *Criminal behavior is learned.* Negatively, this means that criminal behavior is not inherited, as such; also, the person who is not already trained in crime does not invent criminal behavior .... (p. 75)
> *The process of learning criminal behavior by association with criminal and anticriminal patterns involves all of the mechanisms that are involved in any other learning.* Negatively, this means that the learning of criminal behavior is not restricted to the process of imitation. (p. 76)

Cressey (Sutherland & Cressey, 1970, pp. 83-84) then acknowledges that

> in an early period Sutherland states that his theory probably would have to be reviewed to take account of personality traits .... He posed three questions for advocates of *personality traits* as supplements to differential association: (1) What are the personality traits that should be regarded as significant? (2) Are there personality traits, to be used as supplements to differential association, which are not already included in the concept of differential association? (3) Can differential association, which is essentially a *process* of learning, be combined with personal [*sic*] traits, which are essentially the *product* of learning?"

To my mind, Hans Eysenck has answered these questions quite satisfactorily, as we shall see in the next section. Prior to that, let us look at Eysenck's rebuttal of sociological thinking in criminology.

While acknowledging the role of social factors in determining crime, Eysenck (1996, p. 143) pinpoints what, to his mind, are the fallacies of sociological thinking:

1. Sociological theories are not usually put in a way that makes them testable;
2. Arbitrary assumptions are usually made when sociologists are confronted with contradictory evidence.

His conclusion is devastating: "My point is that sociological theories are hunches rather than theories, not based on thorough statistical analysis of historical records, and too inexact to be testable" (p. 144). Elsewhere (Eysenck & Gudjonsson, 1989, p. 4), he suggests that "many—if not most—sociological theories boil down to what might be called *economic* theories, i.e., crime is the result of poverty, whether relative or absolute, deprivation, and similar economic causes. Although such theories have always had an intuitive appeal, they do not accord with the facts." Eysenck and Gudjonsson (1989, p. 12) further argue for a *biosocial* approach, a point also made by Israel Nachshon (1982), who has also criticized sociological theories: "A subjective impression exists that each of these theories contains a basic truth, but only partially" (p. 3). Nachshon reminds his reader of Werner's (1948) organismic-developmental theory, discussed in the previous section. As Nachshon suggests,

> Risking a far-fetched extrapolation of Werner's (1948) developmental model to the history of criminology, it may be conceived that, with the development of the integrative theories, criminology has shifted from the *amorphic* stage of wandering about, through the stage of development of *differential* theories for the explanation of specific concepts of criminal behavior, and now it finally reached the first phases of the third developmental stage, by offering explanations in terms of *integrated* activity of organic, mental and social systems. (p. 4)

As will be seen in the other chapters of this volume, genetic, social and environmental factors must all be considered, in unison with psychological and biological ones, to fully appreciate the field's continuing need for an integrated and integrative vision of the future (see Chapters 4, 5 and 6).

PERSONALITY AND CRIMINALITY REUNITED: THE PSYCHOBIOLOGICAL UNDERPINNING

Eysenck's (1994) psychobiological theory of personality must, by now, be quite familiar to most readers of this book. Developed over the span of 50 years, highly-cited and guiding a programmatic body of research, Eysenckian theory has helped take personality out of the graveyard (with reference to Singer and Kolligan's, 1987, scathing attack on the tradition, in courses on personality, of taking the students through the graveyard of acknowledged masters, without apparent need to look at contemporary developments in the field). Instead of rehashing this material, which is readily available, I prefer to make a number of comments on the theory, as it stands today.

First, the theory presents a multidimensional approach to personality (Eysenck & Eysenck, 1985). The three major (or secondary, or superfactor) traits, extraversion (E), neuroticism (N) and psychoticism (P), are mutually orthogonal dimensions (reconfirmed recently by Glicksohn & Abulafia, 1998), each comprising a number of subtraits (or, primary traits). But there are inconsistencies here. Thus, one finds the primary trait of sensation seeking loading on E, while the trait of impulsivity loads on P. Yet, impulsivity and sensation seeking are sister traits (Zuckerman, 1993): They are positively correlated and have been recently wedded in what

Zuckerman (1994) terms "impulsive sensation seeking." A number of authors, including both Hans Eysenck (1981, 1987, 1993b) and Sybil Eysenck (1993; S. Eysenck & H. Eysenck, 1977), have discussed the relationship between impulsivity and/or sensation seeking and both E and P. The Eysenckian view has, understandably, been that it is better to work at the level of the superfactors than at the level of the primary factors of impulsivity and sensation seeking (Eysenck & Eysenck, 1985, p. 74), and that in fact both traits load on both P and E (and N) (Eysenck, 1994, p. 155). Nevertheless, both the orthogonality and the parsimony of the hierarchical structure become compromised when these inconsistencies are considered. Of course, a major question to consider in planning one's own research is, at what level of the hierarchy should one look for lawfulness? Is it profitable, say, to look at criminality in terms of impulsivity and/or sensation seeking, rather than in terms of extraversion and/or psychoticism? Given that the major scholar of sensation seeking, Marvin Zuckerman (1979, 1991, 1994) is a contributor to this volume, I shall leave it to him to make his own case here (see Chapter 2).

My second point concerning Eysenck's (1994) theory, is that this posits a psycho- or neurobiological underpinning to each personality trait. E is associated with cortical arousal (extraverts being chronically hypoaroused), best tapped using electrophysiological (e.g., Mangan, Pellett, Colrain, & Bates, 1993) and/or functional-imaging (e.g., Johnson *et al.*, 1999) techniques. N is associated with autonomic arousal. P might well be "partly determined by cortical arousal" (Eysenck, 1994, p. 168), though this is an open question, as is the study of P in general (for an overview, see Eysenck, 1992). In fact, given the popularity and proliferation of various multidimensional personality frameworks over the past two decades (Wiggins & Pincus, 1992), one can very well ask just what considerations should be made in discussing and differentiating among the various traits or dimensions of personality. Zuckerman's (1991, 1992) suggestions here are very useful, and I will stress two of particular relevance to the present discussion and to this volume in general:

1. There should be at least moderate heritability for the trait;
2. There should be a clear biological marker or correlate for the trait.

Two chapters in this volume (3 and 4) both address the genetic basis to personality and criminality. Three additional chapters (7, 8 and 9) address various biological markers and neurobiological circuits.

My third comment is that Eysenck's theory has direct implications for the study of criminality and criminal behaviour

(Eysenck & Gudjonsson, 1989), though the predictions are not as tight as one would wish. Extraverts, due to their chronic hypoarousal, are motivated to achieve their optimal level of arousal by increasing the level of stimulation to which they are exposed. This is in accordance with the inverted U curve depicting the relationship between hedonic tone and level of stimulation (Eysenck, 1994, p. 157) and with theories of optimal arousal and their cognitive and motivational implications (Glicksohn, 1992; Hebb, 1955; Zuckerman, 1969, 1979, 1991). The extravert can increase his or her level of arousal by socializing with others (the sociability factor underlying E), by sensation seeking (Zuckerman, 1979) and by exhibiting risky behaviour that might be illegal (Eysenck & Gudjonsson, 1989). Thus, E is a predictor of criminality, and obviously there is a personality x situation interaction, because while all criminals may score high on E, not all those scoring high on E turn to crime (for discussions of environmental and cognitive factors moderating the expression of personality, see chapters 5 and 6 in this volume). Given the change in Eysenckian theory over the years (Eysenck & Eysenck, 1985), with the impulsivity factor previously codetermining E, now being identified with P, there is a second (and complementary) hypothesis in town. That is, that P should be a predictor of criminality (Eysenck, 1992). This brings me to some of the problems revolving around the interpretation of P. As Claridge (1994, p. 149) has recently noted, "the most notable change has been the gradual shift in the item content of the scale, toward questions tapping broadly antisocial, nonconformist traits and away from those of a more manifestly psychotic nature. Indeed, as it now stands the P-scale could be said to have excellent validity as a measure of sociopathic characteristics, a point also emphasized by Zuckerman ... who sees Eysenckian P as overlapping somewhat with his own concept of sensation seeking" (see also Chapter 2).

Thus, while there is a psychobiological underpinning to personality and criminality, there is still much to be worked out, both theoretically and empirically. In this respect, let me note that Depue and Collins (1999) have recently discussed the need to look at the personality dimensions in an interactive sense:

> We are nevertheless in basic disagreement with the positions of Gray, Cloninger, and Zuckerman concerning the lines of causal neurobiological influence, which we propose lie along the two orthogonal dimensions of extraversion and constraint rather than along an impulsivity diagonal resulting

> from their interaction. The main reason for our disagreement is that we believe the impulsivity-sensation seeking traits ... represent *emergent* traits that result from the interaction of extraversion, constraint, and possibly harm-avoidance in the case of sensation seeking ...." (p. 497)
> Secondly, the intrinsically interactive trait of extraversion interacts with neuroticism, entailing the further influence of at least one more neurobiological variable, to form the emergent trait of impulsivity." (p. 497)

The need to look not only at pairwise correlations between traits and phenomena (behaviour, cognition, experience), but also at the effect of the interaction of the traits in determining such phenomena, has been a point stressed previously by Eysenck (1994, p. 158). What Depue and Collins have suggested is that one needs a more dynamic conception of personality, wherein there are emergent traits, such as impulsivity. Such a call for a more holistic, interactive and dynamic view of personality has been advanced by others in the field, such as David Magnusson (e.g., Magnusson & Torestad, 1993; Stattin & Magnusson, 1996). Indeed, one of the themes of the present volume is the need to integrate all of these facets in one's thinking about criminal behaviour. In the next section, I shall present an overview of some of the work being conducted in my own lab, looking at impulsivity and its cognitive manifestations.

TRACKING COGNITION: THE STUDY OF IMPULSIVITY IN THE LAB

Impulsivity may be assessed using trait measures, such as those associated with Barratt (1993, 1994; Barratt & Patton, 1983) and with both H. Eysenck (1987) and S. Eysenck (1993). The construct has also been operationally defined using cognitive (or behavioural) tasks, such as circle tracing (Bachorowski & Newman, 1985, 1990; Kindlon, Mezzacappa & Earls, 1995; Wingrove & Bond, 1997). While most authors agree that the construct is multi-faceted (Barratt, 1993; Barratt & Patton, 1983; Kindlon *et al.*, 1995; White *et al.*, 1994; Wingrove & Bond, 1997), the underlying reason for such a consensus in the literature is that while within each domain (e.g., personality) the tests (or traits) of impulsivity are positively correlated (for personality, see, e.g., Barratt & Patton, 1983;

Zuckerman, 1993), between-domain correlations are negligible (Barratt & Patton, 1983; Wingrove & Bond, 1997). Some headway has been made in looking at time-estimation tasks and their relationship to trait impulsivity (Barratt, 1981, 1985) and to behavioral (cognitive) impulsivity (Bachorowski & Newman, 1985; White *et al.*, 1994; Wingrove & Bond, 1997). Nevertheless, because these latter studies were not guided by considerations taken from within the research area of time perception, their contribution can only be suggestive.

We have employed a series of tasks of time estimation, chosen such that impulsivity could be revealed in deviant performance both within and between tasks (Glicksohn, Aharoni & Segal, in preparation). Assuming that impulsivity is a psychobiological trait (Zuckerman, 1993), and that instructions and cognitive context can moderate but not obliterate the expression of this trait, we hypothesized that impulsivity would be expressed in both slope and intercept values of the psychophysical function expressing perceived duration as a function of objective duration (Glicksohn, 1996, 2001). Conceivably, impulsivity will be more easily apparent using delayed time estimation (Zakay & Fallach, 1984) within a prospective paradigm (Block & Zakay, 1997, because of the infringement on the focus on the immediate "here-and-now" (Damasio's, 1994, "myopia for the future", and Wilson and Herrnstein's, 1985, "present orientation").

In this study, which is still in process, we found as expected (Rule, 1993) that mean slope and mean intercept for delayed prospective reproduction were negatively correlated ($r = -.901$, $p < .0001$). Figure 1 depicts this negative correlation for delayed prospective reproduction ($n = 48$ females), prior to the removal of outliers, with 2 clearly indicated. The first outlier (lower right) scored 89 (maximum for this sample) on The *Barratt Impulsiveness Scale* (BIS-11; Patton, Stanford & Barratt, 1995); in addition, her qualitative score (Q-score, defined here as the square-root of the number of errors) on the Porteus Mazes (Porteus, 1965; Vineland version) was maximal (4.80), and her slope measure was at the minimum (0.27). The second outlier scored 86 on the BIS; her Q-score was 4.12 and her slope measure was 0.37. Clearly, delayed prospective reproduction is a trying task for the impulsive subject. The Q-score correlated positively with BIS ($r = .35$, $p < .05$), and negatively with the slope measure ($r = -.30$, $p < .05$). Note that the Porteus Maze test has been used previously to assess impulsivity, primarily using the Q-score based on errors in style or strategy (Barratt, 1993; Helmers, Young & Pihl, 1995; Lilienfeld, Hess, &

Rowland, 1996; Riddle, & Roberts, 1977; Schalling, & Rosén, 1968). That is, crossing lines, cutting corners, lifting the pencil off from the page, and so forth. Nevertheless, as Fink and McCown (1993) caution, research using this test as a measure of impulsivity has not been uniformly successful. On removing the outliers, these correlations are both reduced in size and become nonsignificant.

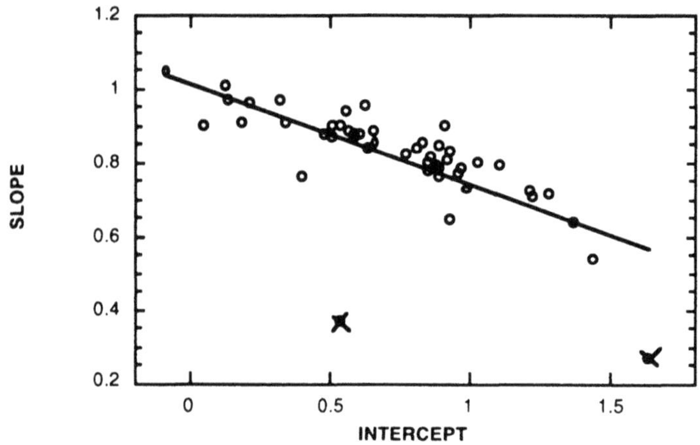

Figure 1. Negative correlation between slope and intercept of the psychophysical function. Two outliers are clearly indicated.

It is also worth mentioning, in bridging to the next section of this chapter, that time perception has been attributed to left-hemisphere cognitive activity (Contreras, Mayagoitia, & Mexicano, 1985; Elbert, Ulrich, Rockstroh, & Lutzenberger, 1991; Polzella, DaPolito, & Hinsman, 1977), there being several speculations regarding the sites involved in time perception. The majority implicate left-frontal activity (Casini & Macar, 1993), left-fronto-central activity (Elbert *et al.*, 1991), or perhaps left-prefrontal activity (Biakofski & Block, 1996; Block, 1996; Casini & Macar, 1999—but see Tracy, Faro, Mohamed, Pinsk, & Pinus, 2000, for a recent dissenting view) or perhaps even right-frontal activity (Gruber, Kleinschmidt, Binkofski, Steinmetz, & von Cramon, 2000; Monfort, Pouthas, & Ragot, 2000; Pouthas, Garnero, Ferrandez, & Renault, 2000). Thus, if impulsivity is revealed in deviant performance in a time-estimation task, then one might need to consider a lateralized (presumably, left-hemispheric) cognitive dysfunction.

THE FUTURE IS HERE: COGNITIVE NEUROSCIENCE AND EXPERIMENTAL CRIMINOLOGY

Writing a decade ago (at the beginning of the "Decade of the Brain"), Nachson (1991, p. 108) noted that he had come to the conclusion a decade previously (Nachshon, 1983—note the change in spelling of his name), "that most of the pertinent evidence points to the possible involvement of the left rather than the right hemisphere in violent behavior. This conclusion was based on a comprehensive review of clinical and experimental evidence gathered by a variety of methods and techniques, which together seem to indicate left-hemisphere dysfunction in psychopathic and violent populations." Some recent work, assessing left-hemisphere dysfunction by means of electroencephalographic (EEG) measures, is generally supportive here (Deckel, Hesselbrock & Bauer, 1996; Pillmann *et al.*, 1999).

Furthermore, the notion of an impairment in interhemispheric exchange was viewed as deserving further consideration (Nachshon, 1983). The latter implicated a corpus-callosal dysfunction, but given no way to feasibly test this directly, such a hypothesis could only be viewed as being speculative (Raine, Buchsbaum & LaCasse, 1997). But with the advent of the use of various neuroimaging modalities in cognitive-neuroscientific research during this past decade, one could actually look for markers of left-hemisphere and corpus-callosal dysfunction *in vivo* and online. Raine *et al.* (1997) compared a group of murderers to suitable controls using positron emission tomography (PET) while the subjects were cognitively engaged. They reported, among other findings, that the murderers exhibited significantly lower glucose metabolism in the corpus callosum. They suggest that such dysfunction, which is indicative of impaired interhemispheric exchange, would imply that the right hemisphere is less regulated by the left (shades of lateral, as opposed to Jacksonian disinhibition?), and that this may (and evidently does) manifest itself as violent behaviour. But I shall leave it to Sharon Ishikawa and Adrian Raine to address this issue elsewhere in this volume (see Chapter 8).

So, imagine how Lombroso would feel, magically transported to the beginning of the 21st century, and being able to scan violent and nonviolent inmates using both structural and functional neuroimaging. Lombroso (1906/1968) had discussed various neuroanatomical aberrations encountered in post-mortem examination of criminal brains, such as "reversed" hemispheric asymmetry and plagiocephaly. With present-day capabilities of magnetic resonance imaging (MRI), one can adequately assess hemispheric asymmetry at the individual level (Glicksohn & Myslobodsky, 1993) and check for "reversed" asymmetry; using computerized tomography (CT) one can adequately assess the incidence of plagiocephaly (Glicksohn, Kapelusshnik, Bar-Ziv & Myslobodsky, 1993; Myslobodsky, Coppola & Weinberger, 1991). Note that in both cases, the assessment is done *in vivo*. One could then move on in one's theorizing, from establishing a case for the prevalence of such peculiarities amongst criminals (this is, note, an open question), to correlating the extent of the neuroanatomic aberration with cognitive dysfunction (e.g., the same left-hemispheric dysfunction, discussed by Nachson). Lombroso would certainly feel at home here, and be able to redress the various attacks on his work and its methodology (for reviews, see Fink, 1985; Wolfgang, 1972). He might even look for further abnormalities, arguing that if (as is commonly believed), criminals have a history of being abused themselves (e.g., Yeager & Lewis, 2000), and/or of perhaps having suffered head injury in the past (e.g., Miller, 1999; Chapter 6), then they might very well exhibit subcortical anomalities (e.g., Myslobodsky *et al.*, 1995). Now imagine Lombroso's face, when introduced to functioning neuroimaging and its use in studying violent behaviour (for a review, see Raine, 1993). He would probably devour such reports (e.g., Raine *et al.*, 1997). This is the promise for this present decade, if cognitive neuroscience and experimental criminology agree to collaborate. The present volume, hopefully, will serve to whet the reader's appetite, and maybe actually encourage such a new phase in research.

This notwithstanding, before embarking on such a scientific quest, one should, however, at least *try* to learn from history. Such theoretical notions as atavism and degeneration fueled *fin de siècle* hysteria regarding the unsavory prospects for society in the twentieth century. Actual historical developments in the twentieth century, such as Nazism and its state-implemented practices of racial cleansing, eugenics and genocide fueled *new* (or *critical*) criminology's abhorrence of neurobiological approaches to criminality. But as Wright and Miller (1998, p. 3) argue, this tendency to "politicize biological explanations of crime, linking these perspectives to sexism, racism, and immoral eugenic policies" helped to foster a biological taboo in the eyes of sociologically-oriented criminologists. Clearly, it would be prudent to divorce scientific thinking (and practice) from pseudoscience, belief and politics. Read on in this volume with an open but critical mind!

REFERENCES

American Psychiatric Association (1994). *DSM-IV: Diagnostic and statistical manual of mental disorders* (4th ed.). Washington, DC: Author.

Andreasen, N.C., Arndt, S., Alliger, R., & Flaum, M. (1995). Symptoms of schizophrenia: Methods, meanings, and mechanisms. *Archives of General Psychiatry*, **52**, 341-351.

Bachorowski, J., & Newman, J.P. (1985). Impulsivity in adults: Motor inhibition and time-interval estimation. *Personality and Individual Differences*, **6**, 133-136.

Bachorowski, J., & Newman, J.P. (1990). Impulsive motor behavior: Effects of personality and goal salience. *Journal of Personality and Social Psychology*, **58**, 512-518.

Barratt, E.S. (1981). Time perception, cortical evoked potentials, and impulsiveness among three groups of adolescents. In J.R. Hays, T.K. Roberts & K.S. Solway (Eds.), *Violence and the violent individual* (pp. 87-95). New York: Spectrum.

Barratt, E.S. (1985). Impulsiveness substrates: Arousal and information processing. In J.T. Spence & C.E. Izard (Eds.), *Motivation, emotion, and personality* (pp. 137-146). New York: Elsevier.

Barratt, E.S. (1993). Impulsivity: Integrating cognitive, behavioral, biological, and environmental data. In W.G. McCown, J.L. Johnson, & M.B. Shure (Eds.), *The impulsive client: Theory, research, and treatment* (pp. 39-56). Washington, DC: American Psychological Association.

Barratt, E.S. (1994). Impulsiveness and aggression. In J. Monahan, & H.J. Steadman (Eds.), *Violence and mental disorder, developments in risk assessment* (pp. 61-79). Chicago: University of Chicago Press.

Barratt, E.S., & Patton, J.H. (1983). Impulsivity: Cognitive, behavioral, and psychophysiological correlates. In M. Zuckerman (Ed.), *The biological bases of sensation seeking, impulsivity, and anxiety* (pp. 77-116). Hillsdale N.J.: Erlbaum.

Biakofski, F., & Block, R.A. (1996). Accelerated time experience after left frontal cortex lesion. *Neurocase* , **2**, 485-493.

Blakney, R.B. (1955). *The way of life (a new translation of the Tao Tê Ching)*. New York: New American Library (Mentor)

Block, R.A. (1996). Psychological time and memory systems of the brain. In J.T. Fraser, & M.P. Soulsby (Eds.), *Dimensions of time and life: The study of time, VIII* (pp. 61-76). Madison, CT: International Universities Press.

Block, R.A., & Zakay, D. (1997). Prospective and retrospective duration judgments: A meta-analytic review. *Psychonomic Bulletin & Review*, **4**, 184-197.

Capruso, D.X. , & Levin, H.S. (1995). Neuropsychiatric aspects of head trauma. In H.I. Kaplan & B.J. Sadock (Eds.), *Comprehensive textbook of psychiatry/VI* (6th ed.) (Vol. 1; pp. 207-220). Baltimore: Williams & Wilkins.

Carlson, E.T. (1985). Medicine and degeneration: Theory and praxis. In J.E. Chamberlin & S.L. Gilman (Eds.), *Degeneration: The dark side of progress* (pp. 121-144). New York: Columbia University Press.

Casini, L., & Macar, F. (1993). Behavioural and electrophysiological evidence for the specific processing of temporal information. *Psychologica Belgia,* **33**, 285-296.

Casini, L., & Macar, F. (1999). Multiple approaches to investigate the existence of an internal clock using attentional resources. *Behavioral Processes,* **45**, 73-85.

Claridge, G. (1994). Psychobiological models and issues. In S. Strack & M. Lorr (Eds.), *Differentiating normal and abnormal personality* (pp. 137-157). New York: Springer.

Contreras, C. M., Mayagoitia, L., & Mexicano, G. (1985). Interhemispheric changes in alpha rhythm related to time perception. *Physiology and Behavior,* **34**, 525-529.

Damasio, A.R. (1994). *Descartes' error: Emotion, reason, and the human brain.* New York: Avon.

Dann, K.T. (1998). *Bright colors falsely seen: Synaesthesia and the search for transcendental knowledge.* New Haven: Yale University Press.

Dennett, D.C., & Kinsbourne, M. (1992). Time and the observer: The where and when of consciousness in the brain. *Behavioral and Brain Sciences,* **15**, 183-247.

Deckel, A.W., Hesselbrock, V., & Bauer, L. (1996). Antisocial personality disorder, childhood delinquency, and frontal brain functioning: EEG and neuropsychological findings. *Journal of Clinical Psychology,* **52**, 639-650.

Depue, R.A., & Collins, P.F. (1999). Neurobiology of the structure of personality: Dopamine, facilitation of incentive motivation, and extraversion. *Behavioral and Brain Sciences,* **22**, 491-569.

Elbert, T., Ulrich, R., Rockstroh, B., & Lutzenberger, W. (1991). The processing of temporal intervals reflected by CNV-like brain potentials. *Psychophysiology,* **28**, 648-655.

Eysenck, H.J. (1981). Book review of M. Zuckerman (1979). *Sensation seeking: Beyond the optimal level of arousal. Contemporary Psychology,* **26**, 187-188.

Eysenck, H.J. (1987). The place of anxiety and impulsivity in a dimensional framework. *Journal of Research in Personality,* **21**, 489-492.

Eysenck, H.J. (1992). The definition and measurement of psychoticism. *Personality and Individual Differences,* **13**, 757-785.

Eysenck, H.J. (1993a). Creativity and personality: Suggestions for a theory. *Psychological Inquiry,* **4**, 147-178.

Eysenck, H.J. (1993b). The nature of impulsivity. In W.G. McCown, J.L. Johnson, & M. B. Shure (Eds.), *The impulsive client: Theory, research, and treatment* (pp. 57-70). Washington, DC: American Psychological Association.

Eysenck, H.J. (1994). Personality: Biological foundations. In P.A. Vernon (Ed.), *The neuropsychology of individual differences* (pp. 151-207). New York: Academic Press

Eysenck, H.J. (1996). Personality and crime: Where do we stand. *Psychology, Crime & Law,* **2**, 143-152.

Eysenck, H.J., & Eysenck, M.W. (1985). *Personality and individual differences: A natural science approach.* New York: Plenum.

Eysenck, H.J., & Gudjonsson, G. (1989). *The causes and cures of criminality.* New York: Plenum.

Eysenck, S.B.G. (1993). The I_7: Development of a measure of impulsivity and its relationship to the superfactors of personality. In W.G. McCown, J.L. Johnson, & M.B. Shure (Eds.), *The impulsive client: Theory, research, and treatment* (pp. 141-149). Washington, DC: American Psychological Association.

Eysenck, S.B.G., & Eysenck, H.J. (1977). The place of impulsiveness in a dimensional system of personality description. *British Journal of Social and Clinical Psychology*, **16**, 57-68.

Fink, A.D., & McCown, W.G. (1993). Impulsivity in children and adolescents: Measurement, causes, and treatment. In W.G. McCown, J.L. Johnson, & M.B. Shure (Eds.), *The impulsive client: Theory, research, and treatment* (pp. 279-308). Washington, DC: American Psychological Association.

Fink, A.E. (1985). *Causes of crime: Biological theories in the Unites States, 1800-1915*. Westport, Conn: Greenwood Press (Originally published in 1938).

Fowles, J. (1970). *The French lieutenant's woman*. London: Panther.

Glicksohn, J. (1992). Subjective time estimation in altered sensory environments. *Environment and Behavior*, **24**, 634-652.

Glicksohn, J. (1995). "Multiple drafts" of subjective experience viewed within a microgenetic framework for cognition and consciousness. *Behavioral and Brain Sciences*, **18**, 807-808.

Glicksohn, J. (1996). Entering trait and context into a cognitive-timer model for time estimation. *Journal of Environmental Psychology*, **16**, 361-370.

Glicksohn, J. (2001). Temporal cognition and the phenomenology of time: A multiplicative function for apparent duration. *Consciousness and Cognition*, **10**, 1-25.

Glicksohn, J., & Abulafia, J. (1998). Embedding sensation seeking within the big three. *Personality and Individual Differences*, **25**, 1085-1099.

Glicksohn, J., Aharoni, R., & Segal, R. (in preparation). Impulsivity and time estimation: Casting a net to catch a fish.

Glicksohn, J., Alon, A., Perlmutter, A., & Purisman, R. (2000-2001). Symbolic and syncretic cognition amongst schizophrenics and visual artists. *Creativity Research Journal*, **13**, 133-143.

Glicksohn, J., Kapelusshnik, J., Bar-Ziv, J., & Myslobodsky, M.S. (1993). A new view on sleep position via head slant. *Functional Neurology*, **8**, 347-350.

Glicksohn, J., & Myslobodsky, M.S. (1993). The representation of patterns of structural brain asymmetry in normal individuals. *Neuropsychologia*, **31**, 145-159.

Gruber, O., Kleinschmidt, A., Binkofski, F., Steinmetz, H., & von Cramon, D.Y. (2000). Cerebral correlates of working memory for temporal information. *NeuroReport*, **11**, 1-5.

Harris, R. (1989). *Murders and madness: Medicine, law, and society in the Fin de Siècle*. Oxford: Clarendon Press.

Hebb, D.O. (1955). Drives and the C.N.S. (conceptual nervous system). *Psychological Review*, **62**, 243-254.

Helmers, K.F., Young, S.N., & Pihl, R.O. (1995). Assessment of measures of impulsivity in healthy male volunteers. *Personality and Individual Differences*, **19**, 927-935.

Jeffery, C.R. (1979). Biology and crime: The new neo-Lombrosians. In C.R. Jeffery (Ed.), *Biology and crime* (pp. 7-18). London: Sage.

Johnson, D.L., Wiebe, J.S., Gold, S.M., Andreasen, N.C., Hichwa, R.D., Watkins, G.L., & Boles Ponto, L.L. (1999). Cerebral blood flow and personality: A positron emission tomography study. *American Journal of Psychiatry*, **156**, 252-257.

Kindlon, D., Mezzacappa, E., & Earls, F. (1995). Psychometric properties of impulsivity measures: Temporal stability, validity and factor structure. *Journal of Child Psychology and Psychiatry*, **36**, 645-661.

Leahey, T.H. (1992). *A history of psychology: Main currents in psychological thought.* Englewood Cliffs, NJ: Prentice Hall.

Lilienfeld, S.O., Hess, T., & Rowland, C. (1996). Psychopathic personality traits and temporal perspective: A test of the short time horizon hypothesis. *Journal of Psychopathology and Behavioral Assessment*, **18**, 285-313.

Lombroso, C. (1901). *The man of genius*. London: Walter Scott.

Lombroso, C. (1968). *Crime: Its causes and remedies* (trans. H.P. Horton). Montclair, NJ: Patterson Smith. (originally published in 1906).

Magnusson, D., & Torestad, B. (1993). A holistic view of personality: A model revisited. *Annual Review of Psychology*, **44**, 427-449.

Mangan, G.L., Pellett, O., Colrain, I.M., & Bates, T.C. (1993). Photic driving and personality. *Personality and Individual Differences*, 15, 329-340.

Martindale, C. (1971). Degeneration, inhibition, and genius. *Journal of the History of the Behavioral Sciences*, **7**, 177-182.

Martindale, C. (1993). Psychoticism, degeneration, and creativity. *Psychological Inquiry*, 4, 209-211.

Martindale, C. (1999). Biological bases of creativity. In R.J. Sternberg (Ed.), *Handbook of creativity* (pp. 137-152). Cambridge: Cambridge University Press.

Miller, E. (1999). Head injury and offending. *Journal of Forensic Psychiatry*, **10**, 157-166.

Monfort, V., Pouthas, V., & Ragot, R. (2000). Role of frontal cortex in memory for duration: An event-related potential study in humans. *Neuroscience Letters*, **286**, 91-94.

Muncie, J., McLaughlin, E., & Langan, M. (1996). (Eds.) *Criminological perspectives: A reader.* London: Sage.

Myslobodsky, M.S., Coppola, R., & Weinberger, D.R. (1991). EEG laterality in the era of structural brain imaging. *Brain Topography*, **3**, 381-390.

Myslobodsky, M.S., Glicksohn, J., Singer, J., Stern, M., Bar-Ziv, J., Friedland, N., & Bleich, A. (1995). Changes of brain anatomy in patients with posttraumatic stress disorder: A pilot magnetic resonance imaging study. *Psychiatry Research: Neuroimaging*, **58**, 259-264.

Nachshon, I. (1982). Toward biosocial approaches in criminology. *Journal of Social and Biological Structures*, **5**, 1-9.

Nachshon, I. (1983). Hemisphere dysfunction in psychopathy and behavior disorders. In M.S. Myslobodsky (Ed.), *Hemisyndromes: Psychobiology, neurology, psychiatry* (pp. 389-414). New York: Academic Press.

Nachson, I. (1991). Neuropsychology of violent behavior: Controversial issues and new developments in the study of hemisphere function. In J.S. Milner (Ed.), *Neuropsychology of aggression* (pp. 93-116). Boston: Kluwer.

Nordau, M. (1895). *Degeneration.* New York: D. Appleton.

Patton, J. H., Stanford, M. S., & Barratt, E. S. (1995). Factor structure of the Barratt impulsiveness scale. *Journal of Clinical Psychology*, **51**, 768-774.

Pillmann, F., Rohde, A., Ullrich, S., Draba, S., Sannemuller, U., & Marneros, A. (1999). Violence, criminal behavior, and the EEG: Significance of left hemispheric focal abnormalities. *Journal of Neuropsychiatry and Clinical Neurosciences*, **11**, 454-457.

Polzella, D.J., DaPolito, F., & Hinsman, M. (1977). Cerebral asymmetry in time perception. *Perception & Psychophysics*, **21**, 187-192.
Porteus, S.D. (1965). *Porteus maze test. Fifty years' application.* Palo Alto: Pacific Books.
Pouthas, V., Garnero, L., Ferrandez, A-M., & Renault, B. (2000). ERPs and PET analysis of time perception: Spatial and temporal brain mapping during visual discrimination tasks. *Human Brain Mapping*, **10**, 49-60.
Raine, A. (1993). *The psychopathology of crime: Criminal behavior as a clinical disorder.* New York: Academic Press.
Raine, A., Buchsbaum, M., & LaCasse, L. (1997). Brain abnormalities in murderers indicated by positron emission tomography. *Biological Psychiatry*, **42**, 495-508
Riddle, M., & Roberts, A.H. (1977). Delinquency, delay of gratification, recidivism, and the Porteus Maze tests. *Psychological Bulletin*, **84**, 417-425.
Rule, S.J. (1993). Analyzing coefficients of psychophysical power functions. *Perception & Psychophysics*, **54**, 439-445.
Schalling, D., & Rosén, A-S. (1968). Porteus Maze differences between psychopathic and non-psychopathic criminals. *British Journal of Social and Clinical Psychology*, **7**, 224-228.
Showalter, E. (1997). *Hystories: Hysterical epidemics and modern culture.* NY: Columbia University Press.
Singer, J.L., & Kolligan, J. (1987). Personality: Developments in the study of private experience. *Annual Review of Psychology*, **38**, 533-574.
Stattin, H., & Magnusson, D. (1996). Antisocial development: A holistic approach. *Development and Psychopathology*, **8**, 617-645.
Sutherland, E.H., & Cressey, D.R. (1970). *Criminology* (8th ed.). Philadelphia: J. B. Lippincott.
Sutherland, E.H., Cressey, D.R., & Luckenbill, D.F. (1992). *Principles of criminology* (11th ed.). Dix Hills, NY: General Hall.
Taylor, J. (Ed.). (1958). *Selected writings of John Hughlings Jackson*, Vol. 2. London: Staples Press.
Tracy, J.I., Faro, S.H., Mohamed, F.B., Pinsk, M., & Pinus, A. (2000). Functional localization of a "time keeper" function separate from attentional resources and task strategy. *NeuroImage* ,**11**, 228-242.
Watson, C.G., Plemel, D., & Burke, M. (1979). Proverb test deficit in schizophrenic and brain-damaged patients. *Journal of Nervous and Mental Disease*, **167**, 561-565.
Werner, H. (1948). *Comparative psychology of mental development.* New York: International Universities Press.
Werner, H. (1978). The concept of development from a comparative and organismic point of view. In S.S. Barten and M.B. Franklin (Eds.), *Developmental processes: Heinz Werner's selected writings*, Volume 1 (pp. 107-130). New York: International Universities Press. (Originally published in 1957)
White, J.L., Moffitt, T.E., Caspi, A., Bartusch, D.J., Needles, D.J., & Stouthamer-Loeber, M. (1994). Measuring impulsivity and examining its relationship to delinquency. *Journal of Abnormal Psychology*, **103**, 192-205.
Wiggins, J.S., & Pincus, A.L. (1992). Personality: Structure and assessment. *Annual Review of Psychology*, **43**, 473-504.
Wilson, J.Q., & Herrnstein, R.J. (1985). *Crime and human nature.* New York: Simon & Schuster.
Wingrove, J., & Bond, A.J. (1997). Impulsivity: A state as well as trait variable. Does mood awareness explain low correlations between trait and behavioural

measures of impulsivity? *Personality and Individual Differences*, **22**, 333-339.

Wolfgang, M.E. (1972). Cesare Lombroso, 1835-1909. In H. Mannheim, (Ed.), *Pioneers in criminology* (2nd ed., pp. 232-291). Montclair, NJ: Patterson Smith.

Wright, D.M. (1975). Impairment in abstract conceptualization in schizophrenia. *Psychological Bulletin*, **82**, 120-127.

Wright, R.A., & Miller, J.M. (1998). Taboo until today? The coverage of biological arguments in criminology textbooks, 1961 to 1970 and 1987 to 1996. *Journal of Criminal Justice*, **26**, 1-19.

Yeager, C.A., & Lewis, D.O. (2000). Mental illness, neuropsychologic deficits, child abuse, and violence. *Child and Adolescent Psychiatric Clinics of North America*, **9**, 793-813.

Zakay, D., & Fallach, E. (1984). Immediate and remote time estimation – a comparison. *Acta Psychologica*, **57**, 69-81.

Zuckerman, M. (1969). Theoretical formulations: 1. In J.P. Zubek (Ed.), *Sensory deprivation: Fifteen years of research* (pp. 407-432). New York: Appleton-Century-Crofts.

Zuckerman, M. (1979). *Sensation seeking: Beyond the optimal level of arousal.* Hillsdale, NJ.: Erlbaum.

Zuckerman, M. (1991). *Psychobiology of personality*. Cambridge, UK: Cambridge University Press.

Zuckerman, M. (1992). What is a basic factor and which factors are basic? Turtles all the way down. *Personality and Individual Differences*, **13**, 675-681.

Zuckerman, M. (1993). Sensation seeking and impulsivity: A marriage of traits made in biology? In W.G. McCown, J.L. Johnson, & M.B. Shure (Eds.), *The impulsive client: Theory, research, and treatment* (pp. 71-91). Washington, DC: American Psychological Association.

Zuckerman, M. (1994). *Behavioral expressions and biosocial bases of sensation seeking.* Cambridge, UK: Cambridge University Press.

PART B

PERSONALITY AND CRIMINALITY: PSYCHOBIOLOGICAL FOUNDATIONS

Chapter 2

PERSONALITY AND PSYCHOPATHY: SHARED BEHAVIORAL AND BIOLOGICAL TRAITS

Marvin Zuckerman, Ph.D.
Department of Psychology, University of Delaware

INTRODUCTION

Evolutionary psychologists might regard the psychopathic personality as the preservation of an adaptive type in the evolutionary history of our species. The "selfish gene" (Dawkins, 1989) may characterize the narcissistic, egocentric psychopath. When the psychopath in the novel "One Flew Over the Cuckoo's Nest" (Kesey, 1964) is asked by a psychiatrist what he thinks the diagnosis of psychopathic personality means, he replies: "Fighting and fucking." Aggressiveness and promiscuity are certainly traits which could be a biological advantage in the promulgation of one's genes. Even today, however, small tribal groups regulate sexual and aggressive expressions, and individuals who transgress the mores of the particular group may be executed or expelled from the group, and this could amount to a death sentence in a hostile environment.

Some members of chimpanzee groups exhibit behaviors resembling those in human psychopaths (Weiss, King, O'Connor, Berry, & Enns, 2001). Normal female chimpanzees show a positive interest in the newborns of other females and touch and hold them if the mother permits. In the group studied by Jane Goodall (1986), a female named "Passion" (ironically?), in collaboration with her adolescent daughter, had the nasty habit of seizing and eating the newborns of other mothers. She was understandably unpopular in the colony and became asocial, spending time only with members of her own family. Cannibalism of infants from other groups has been observed, but chimpanzees do not kill or eat members of their own group. Passion was an aberration in her group, and if chimpanzees used diagnoses she would be called a psychopathic personality. As in many human societies, the most murderous aggression may be practiced against members of other groups, but those who do the same things to members of their own groups are ostracized.

Male chimpanzees will normally mate with any receptive female in estrus, but like humans they seem to have an incest taboo, particularly between mother and son. One chimp observed by Goodall repeatedly tried to rape his mother who was not at all receptive to the advances of her Oedipal son. The same male repeatedly raped or tried to rape his sister and, unlike other males, would attempt to have sex with unreceptive females who were not in estrus. There are limits to acceptable sexual behavior, even in a relatively promiscuous species like chimpanzees.

Murderous emperors and kings, like Caligula and Macbeth, may flourish for a time, but they are often terminated by their peers along with their descendants. More temperate rulers, like Augustus, tend to survive longer and establish lasting dynasties. Evolution may favor dominance and competitive aggression, but the psychopathic personality is not adapted for survival within social groups. Their persistent presence, however, may be due to their procreative proclivities that may have compensated for their other risky behaviors. Life was short and breeding started young among our distant ancestors.

PERSONALITY TRAITS IN PSYCHOPATHY

Psychopathic traits were analyzed into two factors by Harpur, Hare and Hakstian (1989): factor 1 (F1) consists of personality traits like selfishness, egocentricity, callousness, exploitation of others, cruelty, and absence of love, empathy or guilt; factor 2 (F2) contains the behavioral characteristics of the psychopath including social deviance, a parasitic antisocial life style, need for stimulation and proneness to boredom, irresponsibility, and lack of behavioral control or impulsivity. Factor 2 resembles the one identified in factor analyses of personality tests: *Impulsive Unsocialized Sensation Seeking*

(ImpUSS, Zuckerman, Kuhlman, & Camac, 1988; Zuckerman, Kuhlman, Thornquist, & Kiers, 1991).

Harpur *et al.* (1989) correlated ratings of prison inmates on the Psychopathy Check List (PCL) with various personality scales. Table 1 shows selected correlations from this study. The first group of these scales correlates primarily with F2 and the Total PCL, and not with F1 ratings. The MMPI profile for the psychopathic personality includes elevations on two scales: Psychopathic Deviate (Pd), with items describing a pattern of delinquency during youth and current antisocial or egocentric attitudes; the Hypomania (Ma) scale items describe a pattern of high energy and activity. The Socialization scale from the CPI also describes a pattern of delinquency and antisocial behavior and attitudes, but is scored in the prosocial direction so the correlations with the PCL are negative. The Sensation Seeking Scale (SSS) is the operational method of defining the trait of sensation seeking, involving the seeking of novel, intense and complex experiences and the willingness to take risks for the sake of such experience (Zuckerman, 1979, 1994). The Total score and all of the subscales (not shown in this table), correlated significantly with F2 and none of them correlated with F1. The EPQ Psychoticism (P) scale is the best marker for the ImpUSS factor (Zuckerman *et al.*, 1991). P also correlated with F2 but not with F1.

In contrast to the pattern of correlation of P and SS with the Harpur *et al.* PCL ratings, various anxiety scales correlated negatively with F1 but not at all with F2. The EPQ Neuroticism scale correlated negatively with F1 and positively with F2, but the Fenz-Epstein anxiety scale and the KSP Psychic Anxiety scale correlated only with F1.

A distinction is sometimes made between the primary and secondary psychopaths. The primary type, described in Cleckley's (1976) classic work, has antisocial behavior plus the absence of strong emotional responses like anxiety, whereas the secondary type shows the impulsive antisocial behavior, but also has considerable anxiety and tends to be more introverted than the primary type. Emmons and Webb (1974) and Blackburn (1978) distinguished primary and secondary psychopaths using psychometric criteria and contrasted both of these groups with nonpsychopathic prisoners on the SSS. With the exception of the Thrill and Adventure Seeking (TAS) subscale, all of the other subscales distinguished the primary psychopathic from the nonpsychopathic prisoners, but only the Disinhibition subscale showed a significant difference between primary and secondary psychopaths in both studies.

Harpur, Hart and Hare (1994) found that only the Agreeableness scale of the "Big Five" was correlated (negatively) with the PCL ratings in prison inmates, but the facet scores of hostility and excitement seeking were positively related to the PCL. In a student sample, the conscientiousness factor was negatively correlated with the PCL.

Longitudinal studies show that undercontrol or impulsivity, even in children as young as 3 years of age, predicts convictions for violent crimes by age 18 (Henry, Caspi, Moffitt, & Silva, 1996) and diagnosis of Antisocial Personality Disorder (APD), and recidivist criminality at age 26 (Caspi, 2001). Novelty seeking during kindergarten age predicts delinquency at age 13 (Tremblay, Pihl, Vitaro, & Dobkin, 1994).

Table 1. Selected correlations of factor and total score ratings from the Psychopathy Check List (PCL) with personality tests in prisoners.

Personality Tests	n	F1	F2	Total PCL
MMPI Pd	138	.11	.31***	.25**
MMPI Ma	138	.16	.32***	. 27**
MMPI Pd + Ma	138	.18*	.41**	.35***
CPI Socialization	223	-.06	-.44**	-.31***
Sensation Seeking Scale	207	.12	.39***	.30***
EPQ Psychoticism	222	.01	.22***	-.17**
EPQ Extraversion	222	.08	.10	.11
EPQ Neuroticism	222	-.17*	.16*	.02
F-E Anxiety Scale	159	-.17*	.10	-.01
KSP Psychic Anxiety	90	-.39***	-.14	-.30**

*p < .05, ** p<.01, *** p<.001

Note. MMPI = Minnesota Multiphasic Personality Inventory; Pd = Psychopathic Deviate; Ma = Hypomania; CPI = California Personality Inventory; EPQ = Eysenck Personality Questionnaire; F-E = Fenz-Epstein; KSP = Karolinska Scales of Personality

Source: Data from various tables in "Two factor conceptualization of psychopathy: Construct validity and assessment implications," by T. J. Harpur, R. D. Hare, & A. R. Hakstian, 1989, Psychological Assessment, pp. 11-12. Reproduced by permission of the American Psychological Association

Both concurrent and longitudinal studies show that the personality of the psychopath includes strong impulsiveness, sensation seeking, aggression, and weak socialization and behavioral control. In a three factor model these constitute a major dimension of personality which Eysenck and Eysenck (1976) call "Psychoticism." Zuckerman (1989) suggested that a more appropriate clinical label would be "Psychopathy."

DEMOGRAPHICS

Gender

In community studies, about 5 to 7 times as many men as women received a lifetime diagnosis of antisocial personality (Kessler *et al.*, 1994; Robins & Regier, 1991). Every problem associated with conduct disorders in children is more prevalent in boys than in girls, and the differences are particularly large for those involving aggression.

Higher scores for men than for women are also found on the personality traits that are components of the psychopathic personality type: sensation seeking, aggression, impulsivity, and psychoticism. Men are lower on socialization and self-control.

Age

APD can only be diagnosed at age 18 according to the DSM (American Psychiatric Association, 1994) definition, however most cases begin with conduct and oppositional disorders at much earlier ages. Figure 1, based on the data from the community study of Robins and Regier (1991), shows the one year prevalence rates for APD for men and women. The peak for both sexes is in the youngest group (18-29) and declines thereafter, with a very sharp drop in the age group 45-64. This reduction in APD diagnosis in the 50's is reflected in a reduction of their prison time and convictions after 40 years of age (Hare, McPherson, & Forth, 1988).

However, it is the behavioral factor (F2) on the PCL which declines with age. The personality factor (F1) ratings remain level over time (Harpur & Hare, 1994). Sensation seeking and the P scale show a similar age course as APD, peaking in late adolescence and declining with age thereafter (Eysenck & Eysenck, 1976; Zuckerman, 1979, 1994). Beyond age 50, scores on these scales are about half of the level in the 16-19 year-old cohort. Jessor (1983) noted that age trends in personality are "away from involvement in problem behavior and toward commitment to conventional behavior" (p. 329).

When asked to speculate about the source of such gender differences, my female students usually say that they are due to differential socialization of boys and girls, with more latitude allowed for aggression and sensation seeking in boys. Age changes are attributed to the cumulative effect of experience and concomitant increase in

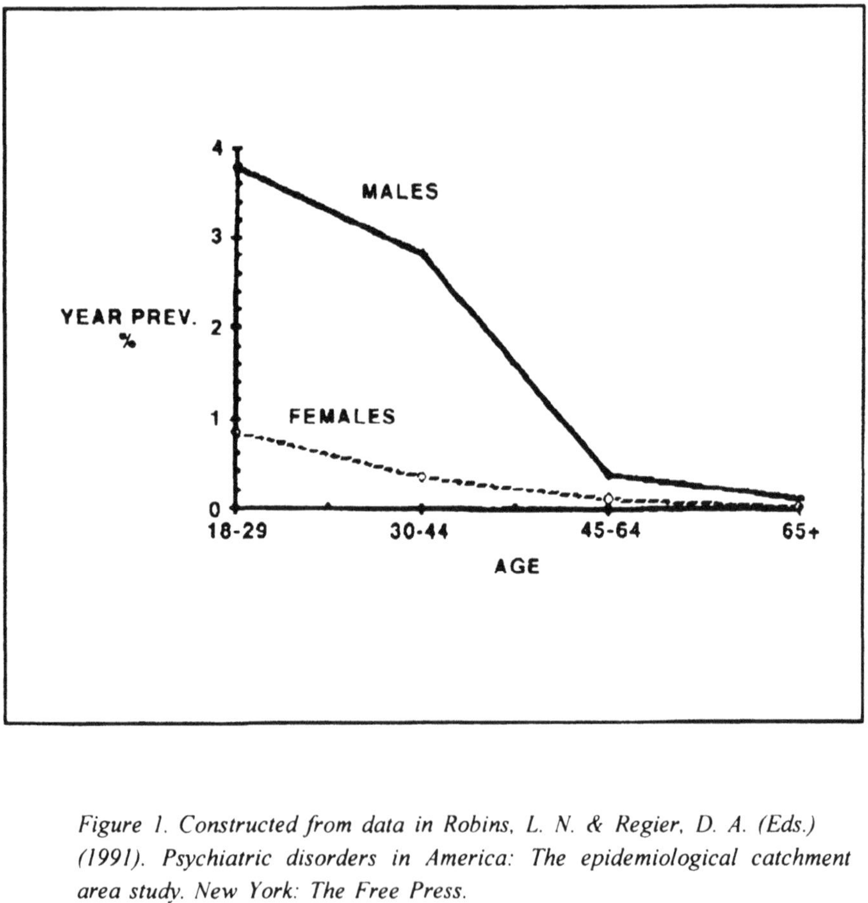

Figure 1. Constructed from data in Robins, L. N. & Regier, D. A. (Eds.) (1991). Psychiatric disorders in America: The epidemiological catchment area study. New York: The Free Press.

"wisdom." It is conceivable, however, that both gender and age differences are due in some part to biological differences and changes. One candidate which could explain the gender and age differences is the hormone testosterone which is higher in men than in women, peaks in adolescence, and declines with age. Another gender- and age-related enzyme is monoamine oxidase. The monoamines themselves show age changes. These will be discussed in relation to psychopathy, aggression, and sensation seeking in a later section.

HEREDITY

The idea of "the bad seed" as an explanation for some types of criminality is anathema to many liberal-minded persons who would put the entire onus of crime on poverty and poor child-rearing. There is, however, considerable evidence from twin and adoption studies of a moderate effect of genetics in chronic criminality (see Chapter 4 in this volume for a detailed discussion). Twin studies of juvenile delinquents show high concordance among both identical and fraternal twins, indicating effects of shared environment and minimal effects of genetics (Cloninger & Gottesman, 1987). In contrast, studies of concordance for criminality in adults show a ratio of concordances for identicals to fraternals of about 2 to 1, indicating moderate genetic effects, and little effect of shared environment. Carey (1992) estimated a heritability of .37 for criminality, assuming imitative influences stronger in identical than fraternal twins, but no influence of shared environment. Lyons (1996) assessed the influences of genetics, shared environment, and unshared environment in twins as a function of phase of life (juvenile vs. adult criminal behavior), and chronicity of criminal behavior (single vs. multiple arrests). Table 2 shows the twin correlations and estimates of genetic and environmental parameters for the different criteria. Before age 15, delinquent criminal behavior showed no significant genetic effects, rather significant effects of shared environment, as well as unshared environment. Early arrests, however, showed significant genetic and shared environmental effects. Later criminal behavior and arrests and felonies, particularly multiple arrests, showed significant, moderate genetic effects, and no effects of shared environment.

Table 2. Twin correlations for arrest and criminality variables and proportions of variance due to additive genetics (G), shared environment (SE), and unique environment (UE).

Variables	ITs (r)	FTs (r)	G	SE	UE
Early criminal behavior	.42	.37	.11	.32***	.58
Early arrest	.73	.53	.39***	.34*	.27
Later criminal behavior	.47	.32	.30*	.17	.53
Later arrested	.45	.30	.30*	.15	.55
Later multiple arrests	.47	.28	.39**	.08	.53
Felony	.59	.40	.38**	.21	.41

*p< .05, **p< .01, ***p< .001

Note. ITs = identical twins; FTs = fraternal twins. Data adapted from "A twin study of self-reported criminal behavior" by M. J. Lyons in Genetics of Criminal and Antisocial Behavior, edited by G. R. Bock & J. A. Goode, 1996. Chichester, UK: Wiley. Copyright 1996. Reproduced by permission of John Wiley & Sons.

Moffitt (1993) suggests that the large group of adolescent delinquents comprises two distinct groups: (1) the *adolescent-limited* group, whose members stop engaging in criminal behavior in their mid-20s, and (2) the *life-course persistent* antisocial group, whose members begin criminal behavior before adolescence and persist long into adult life. The first group is strongly influenced by societal and peer-modeling factors, whereas the second group is relatively more influenced by genetic and non-genetic biological factors influencing neural development and personality traits, such as aggressiveness and impulsivity. This latter group is more likely to fit the diagnostic definition of APD in the DSM. Although smaller in numbers than the larger criminal population, they account for a disproportionate number of crimes.

Can we actually identify the "bad seed" or gene(s)? Molecular genetics now offers that possibility. Ebstein *et al.* (1996) found an association between the gene controlling the development of the dopamine 4 receptor (D4DR) and the personality trait of novelty-seeking (highly correlated with impulsive sensation seeking). The replication record for this finding has been spotty, with some successes and some failures of replication. This particular gene accounts for only 4-5% of the variance in the trait and about 10% of the genetic variance in the trait. Like nearly all traits and forms of psychopathology, this one is polygenetic. However, it is interesting that three studies have shown that the longer form of the gene, which is associated with novelty seeking, is also found in a greater proportion of heroin-dependent subjects than in controls (Ebstein & Belmaker, 1997). Rates of APD among heroin abusers run higher than those of other drug users, ranging from 50 to 70% (Dinwiddie, Reich, & Cloninger, 1992; Rounsaville *et al.*, 1991). The genetic basis for APD may consist of the genes involved in its constituent traits, including sensation seeking (see Chapter 3 for a detailed discussion).

NEUROPSYCHOLOGY

Other contributors to this volume will describe the neuropsychological bases of aggression and psychopathy in more detail (see Chapters 5, 6 and 7). A variety of environmental factors affecting the developing brain could account for brain damage in psychopaths. Bad parenting, involving neglect and poor nutrition, may affect the brain during prenatal and postnatal periods. Birth complications and physical child abuse can also cause brain damage. A study of a Danish cohort found an interaction between birth complications and maternal rejection of infants (Raine, Brennan, & Mednick, 1994). Neither factor alonc predicted an increased criminal record when the infants had reached young adulthood, but the interaction of the two factors revealed an increased rate of violent offenses: 47% of those with both factors became violent criminals, whereas only 20% of those with neither or only one of the two factors had this outcome.

In the classical neuropsychological case of Phineas Gage (Damasio, Grabowski, Frank, Galaburda, & Damasio, 1994), his damaged frontal lobes were assumed to be the source of his change in character from a sober, industrious citizen to a profane, lewd, and disinhibited one. The frontal lobes are regarded as the loci of the executive function of the higher cortical centers, assigning significance to incoming stimuli, integrating plans for action (or restraint), and planning (Luria, 1980).

Gray (1982) describes a pathway from the orbitofrontal cortex to the septum and hippocampus which is involved in detection of signals of punishment or nonreward and inhibition of behavior in response to such signals. If the system is overactive, the result would be anxiety, but if the system is underactive, the individual would be disinhibited. In fact, Gray (1987) labels the psychological trait dimension related to this system: "anxiety-psychopathy." Gorenstein and Newman (1980) describe the "septal syndrome" produced in rats by the lesioning of this system. The resultant behavior consists of an enhanced response for positive reinforcement and a diminished capacity to inhibit responses that have resulted in punishment or loss of reward. Another consequence is enhanced "stimulation seeking." These authors suggest that a deficit in this system, particularly in the frontal cortex, may be a source of the disinhibited behavior of the psychopath.

The arousal hypotheses suggest that as a consequence of a developmental disorder of the brain, psychopaths suffer from a low arousal level or low arousability, and therefore need more intense or novel stimulation to bring them up to an "optimal level of arousal." This might explain the attraction to stimulant drugs like cocaine and amphetamine, but it does not explain the attraction to suppressant drugs like the opiates. The most common type of abnormality found in the EEGs of psychopaths is an excess of slow-wave activity (Syndulko, 1978). Prospective studies have shown that slow-wave activity in the EEGs of boys predicted subsequent theft offenses (Volavka, 1987). Chapter 8 provides a comprehensive review of this literature.

Another form of the arousal hypothesis suggests that psychopaths suffer from a deficit in arousability rather than basal levels of arousal. Cortical evoked potentials (EPs) have been used to test this hypothesis using the response (P300) to an infrequently and unpredictably occurring stimulus (the "odd-ball"). The psychopath is expected to show a weaker P300 response. Although this has not been consistently found in some brain areas, persons diagnosed with APD did show smaller P300 EP amplitudes in frontal areas (O'Connor, Bauer, Tesman, & Hesselbrock, 1994).

Buchsbaum (Buchsbaum, 1971; Buchsbaum & Silverman, 1968) developed an EP paradigm based on the cortical response to a range of stimulus intensities. The contrast between individuals who show an "augmenting" pattern (increasing EP amplitude in response to increasing intensity of stimulation) and those who show a "reducing" pattern (little change of EP amplitude, or decreasing amplitudes of response at the highest intensities of stimulation) has been a source of significant differences. High sensation seekers of the disinhibitory type tend to be EP-augmenters whereas those low on this trait tend to be EP-reducers (Zuckerman, 1990). The results have been replicated many times, particularly for auditory stimuli. Similar relationships have been found for the trait of impulsivity (Carrillo de la Penna & Barratt, 1993). Augmenting has also been found to be the more characteristic pattern in criminals, delinquents, and alcoholics.

The augmenting-reducing paradigm was used in cats (Hall, Rappaport, Hopkins, Griffin, & Silverman, 1970; Lukas & Siegel, 1977; Saxton, Siegel & Lukas, 1987). Augmenter cats were rated as being more exploratory, active, and aggressive, and reducer cats as being more passive and tending to freeze or withdraw in reaction to novel stimuli. Using experimental procedures, Saxton *et al.* (1987) found that augmenter cats learned a fixed interval schedule for positive reinforcement faster than reducers, but that reducers were better on a task requiring a delayed response for reward. In other words, augmenter cats had an enhanced response for reward but a reduced capacity to inhibit response when restraint was necessary to obtain reward. This pattern is precisely that described by Gorenstein and Newman (1980) as the "Septal Syndrome" and thus supports their theory of the role of the orbitofrontal-septohippocampal system in psychopathy.

PSYCHOPHARMACOLOGY

Monoamine Oxidase (MAO)

MAO-Type B is an enzyme involved in the catabolic reduction of monoamines, particularly dopamine (Murphy, Aulakh, Garrick, & Sunderland, 1987). In living humans, MAO-B is obtained from blood platelets. No correlation has been found between MAO-B in platelets and brain, but a host of behavioral (in primates and humans), personality, and psychopathology correlates of MAO obtained from platelets suggest that platelet MAO must index some important brain functions even if only in some specific areas of brain. Swedish investigators have proposed that platelet MAO activity is governed by the same genes that regulate levels of serotonin activity, and that MAO is positively correlated with such activity (Oreland, Wilberg, & Fowler, 1981).

MAO-B is usually negatively correlated with sensation seeking, significantly so in 9 of 13 study groups reviewed by Zuckerman (1994). Low MAO is also found in antisocial and borderline personality disorders (Lidberg, Modlin, Oreland, Tuck, & Gillner, 1985; Reist, Haier, De Met, & Cisz-De Met, 1990) and hyperactivity in children (Shekim *et al.*, 1986). Even in a "normal" population, low MAO is associated with a greater use of tobacco, alcohol and illegal drugs, and convictions for criminal offenses (Coursey, Buchsbaum, & Murphy, 1979).

The relation between MAO and disorders characterized by impulsive sensation seeking, as well as the trait itself, suggests that some of the biological basis for sensation seeking may lie in a dysregulation of monoamine activity. But which of the monoamines is involved? Cloninger, Svrakic and Przybeck (1993) suggest that dopamine activity underlies the trait of novelty seeking. They suggest that low basal levels of activity but high reactivity of the system are characteristic in high novelty seekers. Zuckerman (1991, 1995) proposed that all three monoamines are involved in the supertrait of impulsive unsocialized sensation seeking; with high activity in the dopaminergic system (sensation seeking) system, weak inhibitory activity in the serotonergic system (impulsivity), and low activity in the noradrenergic system (arousal). The combination of systems would produce a person highly susceptible to signals of primary rewards, such as cocaine or sex, inadequately restrained by signals of punishment, and not highly aroused by danger or anxiety, as in the imperturbability of the primary psychopath.

Monoamines

The comparative literature supports these ideas of the functions of the monoaminergic systems in behavior, but the evidence from human studies is sketchy. A blunted serotonin response to an agonist for the transmitter was observed in men diagnosed as having antisocial (Moss, Yao, & Panzak, 1990) and borderline (Siever & Trestman, 1993) personality disorders. The serotonin metabolite, 5-HIAA, in CSF was found to be low in impulsive and alcoholic violent offenders and arsonists (Virkkunen *et al.*, 1994). Conduct and oppositional disorders in children were also associated with low levels of 5-HIAA (Kruesi *et al.*, 1990). Chapter 9 provides a comprehensive review of this literature.

A blunted serotonin response (prolactin) to a serotonin agonist has been found for persons scoring high on the disinhibition sensation seeking scale (Depue, 1995; Netter, Hennig, & Roed, 1996), as well as on the EPQ P scale, and measures of low self-control and impulsivity. No direct relationship was found between an agonist stimulated dopamine response and sensation seeking in these studies. However dopamine response (prolactin) was related to the P scale and to venturesomeness and risk taking from Eysenck's impulsivity scale.

The postulated negative relationship between sensation seeking and the noradrenergic arousal system thus far rests upon one as yet unreplicated finding of a negative correlation between norepinephrine in the CSF and sensation seeking (Ballenger *et al.*, 1983).

Hormones

Testosterone levels are generally related to aggressive behavior, particularly in studies of criminals with histories of violent crimes (Archer, 1994; Archer, Biring & Wu, 1998; Bergman & Brismar, 1994; Dabbs, Carr, Frady, & Riad, 1995). The research relating testosterone to hostility and aggression, as personality traits, in non-prisoners yields more mixed results. Testosterone is influenced by state changes, such as winning or losing in competitive contests. There is a continuous interaction between levels of testosterone and life experiences. Testosterone in socialized individuals is associated with social dominance, assertiveness, sociability, and activity—all positive traits. However, there is also evidence that characteristically high levels of testosterone in less socialized persons makes them more likely to aggress against others. Clinical studies of steroid users show increased aggressiveness in many of them (Pope & Katz, 1994).

Plasma testosterone levels are high in high sensation seekers of the disinhibitory type and average in low sensation seekers (Daitzman & Zuckerman, 1980). Disinhibition is the subscale of the SSS most highly related to psychopathy, and psychopathic behavior is characterized by disinhibited behavior. College fraternities whose member have high levels of testosterone are dedicated to drinking, drugs, disorderly parties, irresponsible sex, and various antisocial activities (Dabbs, 2000). They do not spend much time in prosocial community activities.

High levels of cortisol are associated with stress and depression, but low levels of this hormone are found in prisoners who have a history of psychopathic and violent behavior (Virkkunen, 1985). Low levels of cortisol in a normal population are associated with an unsocialized sensation seeking factor, including the P scale, the SSS disinhibition scale, the MMPI hypomania scale, and lifetime number of sexual partners (Ballenger *et al.*, 1983). Low levels of cortisol in veterans with posttraumatic stress disorder are related to high novelty seeking (Wang, Mason, Charney, & Yehuda, 1997). The low levels of cortisol in criminals or normals may be symptomatic of a lack of stress reactivity or reaction to signals of punishment. High levels of testosterone and low levels of cortisol may have been adaptive in warrior societies from paleolithic times to the present, but in more peaceful societies they may be involved in either normal competitiveness in sports and business or antisocial aggressiveness.

ANIMAL MODELS

In a previous section of this chapter I discussed the augmenting-reducing EP paradigm for disinhibitory sensation seeking and its extension to a behavioral model for sensation seeking in cats. An extension of the paradigm to rats seems to have even more implications for psychopathic personality. Siegel, Sisson and Driscoll (1993) identified the same EP component in rats that was used in augmenting-reducing (A-R) studies of humans and cats, and compared two strains of rats on this psychophysiological trait. The strains were the Roman High-Avoidance (RHA) and Roman Low-Avoidance (RLA) rats. The RHAs were selectively bred for their capacity to learn an active avoidance response, whereas the LHAs were selected for the opposite tendency, that is, they tended to freeze rather than run, and took a longer time to learn the avoidance response. Tested for their EP to different intensities of a visual stimulus, nearly all of the RHA rats showed a sharply increased EP amplitude with increasing stimulus intensity (augmenting), whereas nearly all of the RLA rats demonstrated no change, or reduced EP amplitudes (reducing). Thus a biological marker for sensation seeking, impulsivity and criminality differentiated between the two genetically selected strains of rats.

Since all members of a selectively bred strain have a nearly common set of genes, comparisons on other types of behavior and biological reactions should be informative. Keep in mind that the RHAs are augmenters and the RLAs are reducers, and therefore extrapolating back to the human model, the RHAs would correspond to impulsive sensation seekers and the RLAs to constrained low sensation seekers. Beyond the trait they were bred for (active avoidance vs. behavioral inhibition in response to stress), a number of differences (Table 3) have been found between the two strains.

Comparing the differences between the two strains of rats on behavioral measures with those of high and low impulsive unsocialized sensation seekers at the human level, we see a behavioral similarity in approach or exploration, with little or no fear in novel and potentially risky situations (open-field), a low threshold for aggression, a readiness to use alcohol and tolerance for barbiturates, and nonnuturant maternal behavior. Brain self-stimulation represents the response to intrinsic reward and is a likely basis for substance abuse because nearly every illegal drug releases dopamine from the mesolimbic reward system. The difference between the RHA and RLA is in their response to the intensities of such stimulation. The lows were responsive to low levels of stimulation (reward), but tended to turn-off high levels by pressing a bar, whereas the RHAs responded even more to the high intensities.

Table 3. Key differences between RHA and RLA strains.

Variable	**Main Difference**
Exploration (open-field test)	RHAs more active, less emotionality (defecation).
Shock Induced Aggression	RHAs more and RLAs less aggressive.
Alcohol Drinking	RHAs drink alcohol (in solution), RLAs abstain.
Barbiturates	RHAs high tolerance, RLAs low tolerance.
Maternal Behavior	RHA females less time in nest with young, RLAs more time in nest
Hypothalamic Self-Stimulation (High and low intensities)	RHA less sensitive to high intensity of stimulation, more responsive to high intensities; RLAs more sensitive to low intensities, more escape from high intensities.
Stress Effects (Monoamines and Hormones)	**Main findings**
Prefrontal Cortex	RHAs increased dopamine, RLAs no change in dopamine.
Hypothalamus	RLAs increased serotonin (5HT), RHAs less change in 5HT. RLAs increased corticotropin releasing factor (CRF), RHAs less change in CRF.
Pituitary	RLAs increased adrenocorticotropic hormone (ACTH), RHAs less change in ACTH.

The difference is like that for involuntary cortical reactions in the A-R paradigm. Behaviorally, the psychopath seeks out intense experiences which would be aversive to most non-psychopaths.

I have postulated differences in reactivity of the mesolimbic dopamine system which terminates in the prefrontal cortex. The RHA rats showed an increased reactivity of dopamine in the prefrontal cortex in response to stress. Since the dopamine system regulates movement, the increased release of dopamine in the prefrontal cortex may be related to the behavioral trait for which these rats were bred, namely active avoidance in response to stress.

Serotonergic reactivity is low in psychopathic or other disinhibitory disorders, and the RHA rats show little change in serotonergic release in response to stress in contrast with the RLA rats. The release of serotonin in the RLA rats may be related to their tendency to freeze in stressful situations because the serotonergic system is involved in the inhibition of behavior (Soubrié, 1986) (for a detailed analysis of serotonergic metabolism and aggression, see Chapter 9).

The two strains differ in reactivity of the Hypothalamic-Pituitary-Adrenocortical pathway of stress response. Under stress, the LHA rats show increased CRF and ACTH activity. Presumably, this would result in an increase in release of corticosterone (cortisol in humans). I have cited evidence that psychopathic types and impulsive unsocialized sensation seekers are low in cortisol reactivity.

Another animal model for sensation seeking has been developed by Dellu, Piazza, Mayo, Le Moal and Simon (1996). Rats are selected based on their exploratory reactivity to a novel environment (a circular maze), and designated as high reactive (HR) or low reactive (LR). The differences between these two groups is found in several other types of novel situations. As in humans, the tendency to explore novel situations declines with age in the rats.

When rats were allowed to self-administer amphetamine, the HRs administered more of the drug over a period of 3 days than did the LRs. If deprived of food the HRs eat more in a shorter time, and in an extinction paradigm they have a higher rate of lever press suggesting that they are more sensitive to the results of reward frustration. A salient trait of psychopaths is a lack of "frustration tolerance."

The nucleus accumbens is a primary site for intrinsic reward and the euphoric effects of stimulant drugs. HR rats showed a higher amount of DOPAC (a dopamine metabolite) in the nucleus accumbens on autopsy, indicating a higher basal level of activity in the dopamine transmitter in that nucleus. There was a positive correlation between exploratory activity in response to novelty and the DOPAC content in the nucleus accumbens. Under stress, the dopamine in the nucleus accumbens increased more in the HR than in the LR rats. Basal levels of dopamine activity in the prefrontal cortex (as indicated by a DOPAC/dopamine ratio) were the opposite, with LR rats showing higher activity. The reactivity of dopamine to stress in the prefrontal cortex was not reported, but presumably would be also activated. The authors say that dopamine activity in the prefrontal cortex inhibits activity in the nucleus accumbens.

Both studies support the idea that reactivity of the mesolimbic dopaminergic circuit is the basis of sensation seeking and explains the high susceptibility of sensation seekers and psychopaths to the rewards of drinking and drugs and their intolerance of reward deprivation. Experiments on humans were done in which amphetamine (Hutchison, Wood, & Swift, 1999) and nicotine (Perkins, Wilson, Gerlach, Broge, & Grobe, 2000) were given to high and low sensation or novelty seekers. Sensation and novelty seekers in both cases had more intense subjective effects ("highs") in response to drugs than did low sensation/novelty seekers. Netter *et al.* (1996) found that craving for nicotine was increased by a dopamine agonist more in high than in low sensation seekers. The animal models and the human data suggest that sensation seekers may have a more strongly reactive dopamine system in response to signals of reward and biological reward itself.

SUMMARY AND CONCLUSIONS

Psychopathic Personality describes a personality factor resembling the third major dimension of personality, variously labeled psychoticism, impulsive unsocialized sensation seeking, constraint, and undercontrol. It involves traits of sensation seeking, impulsivity, aggression, asocialization, and lack of self-control and responsibility. Most of these traits show the same relationship to gender and age as does psychopathic personality: higher in men than women, peaking in late adolescence and early adult life and declining with age.

Antisocial behavior first appearing in adolescence (delinquency) shows primarily effects of shared and non-shared environment, but life-course persistent criminality ending in many arrests shows a moderate genetic influence and little or no effect of shared environment. Non-genetic biological factors may also play a role in criminality, particularly violent offenses.

The lack of behavioral inhibition in psychopathy may originate in deviant functioning of the prefrontal cortical and orbitofrontal-septohippocampal system producing an insensitivity to signals of punishment. The psychopath also seems to lack the kind of transmarginal inhibition that dampens responsiveness to intense stimuli.

Another source of the lack of behavioral regulation seems to lie in the low levels of the enzyme MAO that normally regulates activity in monoamine systems, particularly the dopaminergic. Studies of humans and rats suggest that high sensation seekers have a highly reactive dopaminergic system, particularly in the mesolimbic reward system. This may also be true of the psychopathic personality and explains the particular susceptibility of both to drug use and abuse. Low serotonergic system activity also plays a role in the unrestrained behavior of the psychopath.

Testosterone may be part of the explanation for gender and age differences. In unsocialized persons, this hormone contributes to aggressive, violent criminality, but in normal socialized persons it is a source of normal self-assertion, competitiveness, and socialized sensation seeking.

Animal models, using biological markers for traits like sensation seeking in humans to identify analogous traits in other species, have been useful in more direct testing of the neuropharmacological hypotheses about these traits.

Psychopathy as a psychiatric disorder appears relatively late in medical history, in the nineteenth century, with diagnoses like "moral insanity." The medicalization of a disorder always implies that there are biological factors involved in the etiology. Even in the time of Hippocrates, it was recognized that brain injuries could cause changes in temperament (Roccatagliata, 1986).

The "bad seed" idea became identified with reactionary ideologies that denied all involvement of social forces like poverty in criminal behavior. However, the dissociation of the group of criminals called psychopathic from the larger body of criminals allowed a closer examination of this group who account for such a disproportionate amount of crime. The roots of psychopathy involve an interaction between genetic vulnerability, birth complications, and parental rejection. We might amend the "bad seed" idea to: bad seed in bad soil with inadequate sun and rain. Weeds thrive and grass dies in such conditions.

REFERENCES

American Psychiatric Association (1994). *Diagnostic and statistical manual of mental disorders* (4th ed.). Washington, DC: Author

Archer, J. (1994). Testosterone and aggression. *Journal of Offender Rehabilitation*, **21**, 3-39.

Archer, J., Biring, S.S., & Wu, F.C. (1998). The association between testosterone and aggression in young men: Empirical findings and a meta-analysis. *Aggressive Behavior*, **24**, 411-420.

Ballenger, J.C., Post, R.M., Jimerson, D.C., Lake, C.R., Murphy, D.L., Zuckerman, M., & Cronin, C. (1983). Biochemical correlates of personality traits in normals: An exploratory study. *Personality and Individual Differences*, **4**, 615-625.

Bergman, B., & Brismar, B. (1994). Hormone levels and personality traits in abusive male alcoholics. *Alcoholism: Clinical and Experimental Research*, **18**, 311-316.

Blackburn, R. (1978). Electrodermal and cardiovascular correlates of psychopathy. In R. D. Hare & D. Schalling (Eds.), *Psychopathic behavior: Approaches to research* (pp. 157-164). New York: Wiley.

Buchsbaum, M.S. (1971). Neural events and the psychophysical law. *Science*, **172**, 502.

Buchsbaum, M.S., & Silverman, J. (1968). Stimulus intensity control and the cortical evoked response. *Psychosomatic Medicine*, **30**, 12-22.

Carey, G. (1992). Twin imitation for antisocial behavior: Implications for genetic and family environment research. *Journal of Abnormal Psychology*, **101**, 18-25.

Carrillo de la Penna, M.T., & Barratt, E.S. (1993) Impulsivity and ERP augmenting/reducing. *Personality and Individual Differences*, **15**, 25-32.

Caspi, A. (2001). *How early can we tell? Personality continuity from childhood to adulthood.* Paper presented at 10th Biennial meeting of the International Society for the Study of Individual Differences. Edinburgh, Scotland.

Cleckley, H. (1976). *The mask of sanity* (5th ed.). St. Louis, MO: Mosby.

Cloninger, C.R., & Gottesman, I.I. (1987). Genetic and environmental factors in antisocial behavior. In S.A. Mednick, T.E. Moffitt, & S.A. Stack (Eds.), *The causes of crime: New biological approaches* (pp. 92-109). Cambridge: Cambridge University Press.

Cloninger, C.R., Svrakic, D.M., & Przybeck, T.R. (1993). A psychobiological model of temperament and character. *Archives of General Psychiatry*, **50**, 975-990.

Coursey, R.D., Buchsbaum, M.S., & Murphy, D.L. (1979). Platelet MAO activity and evoked potentials in the identification of subjects biologically at risk for psychiatric disorders. *British Journal of Psychiatry*, **134**, 372-381.

Dabbs, J.M. (2000). *Heroes, rogues, and lovers*. New York: McGraw-Hill.

Dabbs, J.M.,Jr.,Carr, T.S., Frady, R.L., & Riad, J.K. (1995). Testosterone, crime, and misbehavior among 692 male prison inmates. *Personality and Individual Differences*, **18**, 627-633.

Daitzman, R., & Zuckerman, M. (1980). Disinhibitory sensation seeking, personality and gonadal hormones. *Personality and Individual Differences*, **1**, 103-110.

Damasio, H., Grabowski, T., Frank, R., Galaburda, A.M., & Damasio, A.R. (1994). The return of Phineas Gage: Clues about the brain from the skull of a famous patient. *Science*, **264**, 1102-1105.

Dawkins, R. (1989). *The selfish gene* (2nd ed.). Oxford: Oxford University Press.

Dellu, F., Piazza, P.V., Mayo, W., Le Moal, M., & Simon, H. (1996). Novelty seeking in rats: Biobehavioral characteristics and possible relationships with the sensation-seeking trait in man. *Neuropsychobiology*, **34**, 136-145.

Depue, R.A. (1995). Neurobiological factors in personality and depression. *European Journal of Personality*, **9**, 413-439.

Dinwiddie, S.H., Reich, T., & Cloninger, C.R. (1992). Psychiatric comorbidity and suicidality among intravenous drug users. *Journal of Clinical Psychiatry*, **53**, 364-369.

Ebstein, R.P., & Belmaker, R.H. (1997). Saga of an adventure gene: Novelty seeking, substance abuse and the dopamine D4 receptor (D4DR) exon III repeat polymorphism. *Molecular Psychiatry*, **2**, 381-384.

Ebstein, R.P., Novick, O., Umansky, R., Priel, B., Osher, Y., Blaine, D., Bennett, E.R., Nemanov, L., Katz, M., & Belmaker, R.H. (1996). Dopamine D4 receptor (D4DR) exon III polymorphism associated with the human personality trait of novelty seeking. *Nature Genetics*, **12**, 78-80.

Emmons, T.D., & Webb, W.W. (1974). Subjective correlates of emotional responsivity and stimulation seeking in psychopaths, normals and acting-out neurotics. *Journal of Consulting and Clinical Psychology*, **42**, 620-625.

Eysenck, H.J., & Eysenck, S.B.G. (1976). *Psychoticism as a dimension of personality*. London: Hodder & Stoughton.

Goodall, J. (1986). *The chimpanzees of Gombe*. Cambridge, MA: The Belknap Press, Harvard University Press.

Gorenstein, E.E., & Newman, J.P. (1980). Disinhibitory psychopathology: A new perspective and a model for research. *Psychological Review*, **87**, 301-315.

Gray, J.A. (1982). *The neuropsychology of anxiety: An enquiry into the functions of the septohippocampal system*. New York: Oxford University Press.

Gray, J.A. (1987). The neuropsychology of emotion and personality. In S.M. Stahl, S.D. Iverson, & E.C. Goodman (Eds.), *Cognitive neurochemistry* (pp. 171-190). Oxford: Oxford University Press.

Hall, R.A., Rappaport, M., Hopkins, H.K., Griffin, R.B., & Silverman, J. (1970). Evoked response and behavior in cats. *Science*, **170**, 998-1000.

Hare, R.D., McPherson, L.M., & Forth, A.E. (1988). Male psychopaths and their criminal careers. *Journal of Consulting and Clinical Psychology*, **56**, 710-714.

Harpur, T.J., & Hare, R.D. (1994). Assessment of psychopathy as a function of age. *Journal of Abnormal Psychology*, **103**, 604-609.

Harpur, T.J., Hare, R.D., & Hakstian, A.R. (1989). Two-factor conceptualization of psychopathy: Construct validity and assessment implications. *Psychological Assessment*, **1**, 6-17.

Harpur, T.J., Hart, S.D., & Hare, R.D. (1994). The personality of the psychopath. In P.T. Costa, & T.A. Widiger (Eds.), *Personality disorders and the five-factor model of personality* (pp. 198-216). Washington, DC: American Psychological Association.

Henry, B, Caspi, A., Moffitt, T.E., & Silva, P.A. (1996). Temperamental and familial predictors of violent and nonviolent criminal convictions: Age 3 to age 18. *Developmental Psychology*, **32**, 614-623.

Hutchison, K.E., Wood, M.D., & Swift, R. (1999). Personality factors moderate subjective and psychophysiological responses to d-amphetamine in humans. *Experimental and Clinical Psychopharmacology*, **7**, 493-501.

Jessor, R. (1993). The stability of change: Psychosocial development from adolescence to young adulthood. In D. Magnusson, & V. Allen (Eds.),

Human development: An interactional perspective (pp. 321-341). New York: Academic Press.

Kesey, K. (1964). *One flew over the cuckoo's nest.* New York: Viking Press.

Kessler, R.C., McGonagle, K.A., Zhao, S., Nelson, C.B., Hughes, M., Eshleman, S, Wittchen, H-U., & Kendler, K.S. (1994). Lifetime and 12-month prevalence of DSM-III-R psychiatric disorders in the United States. *Archives of General Psychiatry*, **51**, 8-19.

Kruesi, M.J.P., Rapoport, J.L., Hamburger, S., Hibbs, E., Potter, W.Z., Lemane, M., & Brown, G.L. (1990). Cerebrospinal fluid monoamine metabolites, aggression, and impulsivity in disruptive behavior disorders of children and adolescents. *Archives of General Psychiatry*, **47**, 419-426.

Lidberg, L., Modlin, I., Oreland, L., Tuck, J.R., & Gillner, A. (1985). Platelet monoamine oxidase and psychopathy. *Psychiatry Research*, **16**, 339-343.

Lukas, J.H., & Siegel, J. (1977). Cortical mechanisms that augment or reduce evoked potentials in cats. *Science*, **196**, 73-75.

Luria, A.R. (1980). *Higher cortical function in man* (2nd ed.). New York: Basic Books.

Lyons, M.J. (1996). A twin study of self-reported criminal behaviour. In G.R. Bock & J.A. Goode (Eds.), *Genetics of criminal and antisocial behaviours* (pp. 61-70). Chichester: Wiley.

Moffitt, T.E. (1993). Adolescence-limited and life course persistent antisocial behavior: A developmental taxonomy. *Psychological Review*, **100**, 674-701.

Moss, H.B., Yao, J.K., & Panzak, G.L. (1990). Serotonergic responsivity and behavioral dimensions in antisocial personality disorder with substance abuse. *Biological Psychiatry*, **28**, 325-338.

Murphy, D.L., Aulakh, C.S., Garrick, N.A., & Sunderland, T. (1987). Monoamine oxidase inhibitors as antidepressants: Implications for the mechanism of action of antidepressants and the psychology of the affective disorders and some related disorders. In H.Y. Meltzer (Ed.), *Psychopharmacology: The third generation of progress* (pp. 545-552). New York: Raven Press.

Netter, P., Hennig, J., & Roed, I.S. (1996). Serotonin and dopamine as mediators of sensation seeking behavior. *Neuropsychobiology*, **34**, 155-165.

O'Connor, S., Bauer, L., Tesman, A., & Hesselbrock, V. (1994). Reduced P3 amplitudes are associated with both a family history of alcoholism and antisocial personality disorder. *Progress in Neuro-Psychopharmacology and Biological Psychiatry*, **18**, 1307-1321.

Oreland, L., Wilberg, A., & Fowler, C.J. (1981). Monoamine oxidase activity as related to monoamine oxidase activity and monoaminergic function in the brain. In B. Angrist (Ed.), *Recent advances in neuropsychopharmacology* (Vol. 31). Oxford: Pergamon.

Perkins, K.A., Wilson, A., Gerlach, D., Broge, M., & Grobe, E. (2000). Greater sensitivity to subjective effects of nicotine in nonsmokers high in sensation seeking. *Experimental and Clinical Psychopharmacology*, **8**, 462-471.

Pope, H.G., & Katz, D.L. (1994). Psychiatric and medical effects of anabolic-androgenic steroid use: A controlled study of 160 athletes. *Archives of General Psychiatry*, **51**, 375-382.

Raine, A., Brennan, P., & Mednick, S.A. (1994). Birth complications combined with early maternal rejection at 1 year predispose to violent crime at age 18 years. *Archives of General Psychiatry*, **51**, 984-988.

Reist, C., Haier, R.J., De Met, E., & Cicz-De Met, A. (1990). Platelet MAO activity in personality disorders and normal controls. *Psychiatry Research*, **30**, 221-227.

Robins, L.N., & Regier, D.A. (Eds.) (1991). *Psychiatric disorders in America: The epidemiological catchment area study.* New York: The Free Press.

Roccatagliata, G. (1986). *A history of ancient psychiatry.* New York: Greenwood Press.

Rounsaville, B.J., Anton, S.F., Carroll, K., Budde, D., Prusoff, D., & Gavin, F. (1991). Psychiatric diagnosis of treatment seeking cocaine abusers. *Archives of General Psychiatry*, **48**, 43-51.

Saxton, P.M., Siegel, J., & Lukas, J.H. (1987). Visual evoked potential augmenting-reducing slopes in cats 2: Correlations with behavior. *Personality and Individual Differences*, **8**, 511-519.

Shekim, W.O., Bylund, D.B., Alexson, J., Glaser, R.D., Jones, S.C., Hodges, K., & Perdue, S. (1986). Platelet MAO and measures of attention and impulsivity in boys with attention deficit disorder and hyperactivity. *Psychiatric Research*, **18**, 179-188.

Siegel, J., Sisson, D.F., & Driscoll, P. (1993). Augmenting and reducing of visual evoked potentials in Roman high- and low-avoidance rats. *Physiology and Behavior*, **54**, 707-711.

Siever, L., & Trestman, R.L. (1993). The serotonin system and aggressive personality disorder. *International Journal of Clinical Psychopharmacology*, **8**, 33-39.

Soubrié, P. (1986). Reconciling the role of central serotonin neurons in human and animal behavior. *Behavioral and Brain Sciences*, **9**, 319-364.

Syndulko, K. (1978). Electrocortical investigations of sociopathy. In R. D. Hare & D. Schalling (Eds.), *Psychopathic behaviour: Approaches to research* (pp. 145-156). Chichester: Wiley.

Tremblay, R.E., Pihl, R.O., Vitaro, F., & Dobkin, P.L. (1994). Predicting early onset of male antisocial behavior from preschool behavior. *Archives of General Psychiatry*, **51**, 732-739.

Virkkunen, M. (1985). Urinary free cortisol secretion in habitually violent offenders. *Acta Psychiatrica Scandinavica*, **72**, 40-44.

Virkkunen, M., Rawlings, R., Tokola, R., Poland, R. E., Guidotti, A., Nemeroff, C., & Bissette, G. (1994). CSF biochemistries, glucose metabolism, and diurnal activity in alcoholic, violent offenders, fire setters, and health volunteers. *Archives of General Psychiatry*, **51**, 20-27.

Volavka, J. (1987). Electroencephalogram among criminals. In S.A. Mednick, T.E. Moffitt, & S.A. Stack (Eds.), *The causes of crime: New biological approaches* (pp. 137-145). Cambridge: Cambridge University Press.

Wang, S., Mason, J., Charney, D., & Yehuda, R. (1997). Relationships between hormonal profile and novelty seeking in combat-related posttraumatic stress disorder. *Biological Psychiatry*, **41**, 145-151.

Weiss, A., King, J.E., O'Connor, L.E., Berry, J.W., & Enns, R.M. (2001). *Assessing the construct of antisocial personality disorder in chimpanzees (Pan troglodytes).* Paper presented at 10th Biennial meeting of the International Society for the Study of Individual Differences. Edinburgh, Scotland.

Zuckerman, M. (1979). *Sensation seeking: Beyond the optimal level of arousal.* Hillsdale, NJ: Erlbaum.

Zuckerman, M. (1989). Personality in the third dimension: A psychobiological approach. *Personality and Individual Differences*, **10**, 391-418.

Zuckerman, M. (1990). The psychophysiology of sensation seeking. *Journal of Personality*, **58**, 313-345.

Zuckerman, M. (1991). *Psychobiology of personality.* Cambridge, UK: Cambridge University Press.

Zuckerman, M. (1994). *Behavioral expressions and biosocial bases of sensation seeking*. New York: Cambridge University Press.

Zuckerman, M. (1995). Good and bad humors: Biochemical bases of personality and its disorders. *Psychological Science*, **6**, 325-332.

Zuckerman, M., Kuhlman, D.M., & Camac, C. (1988). What lies beyond E and N? Factor analyses of scales believed to measure basic dimensions of personality. *Journal of Personality and Social Psychology*, **54**, 96-107.

Zuckerman, M., Kuhlman, D., Thornquist, M., & Kiers, H. (1991). Five (or three) robust questionnaire scale factors of personality without culture. *Personality and Individual Differences*, **12**, 929-941.

Chapter 3

GENETICS OF SENSATION OR NOVELTY SEEKING AND CRIMINAL BEHAVIOR

Richard P. Ebstein, M.D. and Robert H. Belmaker, M.D.
Herzog Hospital, Jerusalem, and Faculty of Health Sciences, Ben-Gurion University of the Negev

INTRODUCTION

Criminality is a complex phenotype and is characterized by a wide spectrum of acts, from non-violent white collar crimes to serial killings. The focus of this review is criminal behavior associated with acts of violence, and we will review the evidence that violent behavior is correlated with heritable personality dimensions, particularly Novelty or Sensation Seeking. For a panoramic view of Sensation Seeking, see Chapter 2 of this volume; Chapter 4 further develops the relationship between genetics and crime.

Individuals who show so-called antisocial behaviors are recognized by DSM IV criteria as people who do not conform to social norms and exhibit Antisocial Personality Disorder (ASPD) (Kaplan, Sadock, & Grebb, 1994). The prevalence is 3% in men and 1% in woman. The onset of the disorder is before the age of 15. As many as 75% of some prisoner cohorts may be diagnosed with ASPD. These individuals are characterized as being tense, hostile, irritable and full of rage. Specific diagnostic criteria include, among others, such items as impulsivity, irritability and aggressiveness as indicated by repeated physical fights or assaults, reckless disregard for safety of self or others, and so forth. There is evidence for a familial occurrence of ASPD (Nigg & Goldsmith, 1994) that appears to be largely due to genetic factors rather than to shared environmental factors. A large-scale twin study of Vietnam veterans showed that for adults with ASPD, genetic factors were predominant (Lyons *et al.*, 1995).

Novelty or Sensation Seeking: The Dark Side of a Personality Trait

ASPD is a psychopathology presenting as an extreme manifestation of normal personality dimensions. Individuals diagnosed with one of the DSM IV Personality Disorder clusters do not perceive of themselves as being ill, but regard their 'disorder' as an essential component of their personalities—ASPD essentially describes who they are. One of the conspicuous features of this psychopathic behavior is Novelty or Sensation Seeking. For example, in a study of parolees, laboratory evaluations of Impulsivity (a prominent feature in Novelty or Sensation Seeking personality traits—see discussion below) significantly correlated with violent and aggressive behavior (Cherek, Moeller, Dougherty, & Rhoades, 1997). In another investigation (Hare & McPherson, 1984), Impulsivity (one of 22 items in the author's Psychopathy Check List or PCL) was the most highly correlated item with a discriminant function used to classify prison inmates into violent and non-violent groups. Proneness to Boredom also correlated highly with violent behavior. Both of these personality features are principal features of Novelty or Sensation Seeking behavior.

Self-report questionnaires developed by psychologists over the past three decades identify a personality factor (Novelty or Sensation Seeking) which reasonably correlates among tests (Grosz *et al.*, 1994; McCourt, Gurrera, & Cutter, 1993; von Knorring, von Knorring, Smigan, Lindberg, & Edholm, 1987; Zuckerman, 1994) including Cloninger's *Novelty Seeking* (Cloninger, 1986), Zuckerman's *Impulsivity-Sensation Seeking* (Zuckerman, 1994), Eysenck's *Psychoticism* and *Extraversion* (Eysenck, 1998) and some of the Big Five factors inventoried by the NEO personality instrument (McCrae & Costa, 1997). This personality dimension is often, but not necessarily, associated with maladaptive behaviors; Zuckerman's Impulsive-Sensation Seeking factors correlate with substance abuse (La Grange, Jones, Erb, & Reyes, 1995), smoking (Carton, Jouvent, & Widlocher, 1994), risky driving especially while intoxicated (Jonah, 1997), 'unsafe sex' (Schroth, 1996), injury (Mawson *et al.*, 1996), marital instability (Ficher, Zuckerman, & Neeb, 1981), criminality

(Young, 1990), ASPD (Blackburn, 1987) and aggression (Zuckerman, 1994).

Novelty Seeking is one of four temperament factors defined by Cloninger's Tridimensional Personality Questionnaire. This factor comprises elements of impulsivity, extraversion, exploratory activity, curiosity, and disorderliness, and correlates highly with Zuckerman's Sensation Seeking dimension. Alcoholics are distinguished by high Novelty Seeking scores, especially Cloninger's type II variety characterized by early onset and family history (Wills, Windle, & Cleary, 1998). Novelty Seeking also figures prominently in other varieties of substance abuse, including opiate abuse (Vukov, Baba-Milkic, Lecic, Mijalkovic, & Marinkovic, 1995) and cigarette smoking (Pomerleau, Pomerleau, Flessland, & Basson, 1992). We have observed that Israeli heroin addicts score significantly higher on one subscale (NS3) of the Novelty Seeking dimension (Mel *et al.*, 1998).

Similar relationships were observed when investigators employed Eysenck's Personality Questionnaire. Smoking was associated with Extraversion and Psychoticism (Arai, Hosokawa, Fukao, Izumi, & Hisamichi, 1997). Heightened impulsivity and extraversion are associated with degree of severity of psychological and behavioral change in pathological gamblers (Blaszczynski, Steel, & McConaghy, 1997). Alcoholics score higher on Psychoticism (Cruz *et al.*, 1995). Abstinence after one year corresponded negatively with impulsivity in alcoholics (McCown, 1989). Again, risky driving correlates with Extraversion and Psychoticism (Martin & Boomsma, 1989). Factor analysis of Eysenck's psychoticism dimension demonstrated that impulsiveness is an important trait in Psychoticism (P) (Howarth, 1986).

Novelty or Sensation Seeking: The Bright Side of a Personality Trait

Novelty or Sensation Seeking, similar to the other TPQ temperament factors, is also part of the normal repertoire of human personality, with many positive facets. For example, Air Force pilot recruits had higher scores in sensation seeking scales suggesting disinhibited behavior in the social sphere, interest in sports and activities involving some danger, and a need for change. They also had higher scores on an impulsivity scale that comprises sensation seeking content (Klinteberg *et al.*, 1992). Not surprisingly, similar high sensation seeking has been observed in people engaged in some sports, such as bungy jumping (Michel, Carton, & Jouvent, 1997) and sky diving (Hymbaugh & Garrett, 1974).

GENETIC FACTORS IN DETERMINING PERSONALITY TRAITS RELATED TO NOVELTY AND SENSATION SEEKING

Heritability is an important determinant of personality factors (Bouchard, 1994) including the Novelty or Sensation Seeking trait. The genetic architecture of Sensation Seeking was analyzed in 1591 adolescent twin pairs, and a simple additive genetic model best-explained individual differences. Between 48 and 63% of the total variance was attributable to genetic factors (Koopmans, Boomsma, Heath, & van Doornen, 1995). Not unexpectedly, since a number of studies have demonstrated associations between sensation seeking traits and measures of impulsivity, a genetic correlation was also observed between these two traits (Hur & Bouchard, 1997). Similarly, Novelty Seeking and the other traits in Cloninger's TPQ show strong heritability of 50-60% (Heath, Cloninger, & Martin, 1994). Genetic factors also account for a significant percentage of the variance in accounting for individual differences in Eysenck's personality traits of Neuroticism, Extraversion and Psychoticism (Floderus-Myrhed, Pedersen, & Rasmuson, 1980).

Has a Gene been Found Contributing to Novelty or Sensation Seeking?

Although the evidence is strong that heritable factors account for as much as 50% of individual differences in assessment of temperament factors, only recently have specific genes been linked to particular personality traits (Benjamin *et al.*, 1996; Ebstein *et al.*, 1996). Our seminal study catalyzed many investigations by others and ourselves aimed at validating these first reports, and identifying other genes that contribute to personality factors.

These two initial studies examined the role of the exon III 48 bp repeat dopamine D4 receptor polymorphism in personality. This gene has a highly polymorphic 16 amino acid repeat region in the putative third cytoplasmic loop that varies between 2-10 repeats in most populations and changes the length of the receptor protein (Lichter *et al.*, 1993; Van Tol, 1996, 1998). There is some evidence that the long and short forms of this protein have a moderate functional significance (Asghari *et al.*, 1994, 1995), although a more recent study showed that the polymorphic region of the loop does not appear to affect the specificity or efficiency of G(i)alpha coupling (Kazmi, Snyder, Cypess, Graber, & Sakmar, 2000).

Two approaches were used. The Israeli study was a case-control design, whereas the NIH group's sample also included a number of sib pairs allowing for both a between- and a within-families design. Both studies suggested an association between Novelty Seeking (NS) and the long alleles of the DRD4 receptor. Although some subsequent studies confirmed our initial findings (Bau *et al.*, 2001; Kuhn *et al.*, 1999; Noble *et al.*, 1998; Ono *et al.*, 1997; Strobel, Wehr, Michel, & Brocke, 1999; Tomitaka *et al.*, 1999), others did not (Bau, Roman, Almeida, & Hutz, 1999; Gebhardt *et al.*, 2000; Gelernter *et al.*, 1997; Herbst, Zonderman, McCrae, & Costa, 2000; Jonsson *et al.*, 1998; Malhotra *et al.*, 1996; Persson *et al.*, 2000; Pogue-Geile, Ferrell, Deka, Debski, & Manuck, 1998; Sander *et al.*, 1997; Sullivan *et al.*, 1998; Vandenbergh, Zonderman, Wang, Uhl, & Costa, 1997).

We note, however, that three recent studies appear to confirm the relevancy of the DRD4 gene to NS. A Japanese group has characterized several polymorphisms in the upstream promoter region of the DRD4 gene, and one of these was associated with higher NS scores in a group of Japanese subjects (Okuyama *et al.*, 2000). This particular promoter region polymorphism is functionally significant and the T variant of the C-521T polymorphism reduces transcriptional efficiency. This association between the promoter region polymorphism and Novelty Seeking was replicated in a Hungarian population (Ronai *et al.*, 2001) but not in a second Japanese group (Mitsuyasu *et al.*, 2001). In a second study, a Finnish group has shown an association between the DRD4 exon III 2 and 5 repeat alleles and high NS scores in a very large and homogeneous Finnish birth cohort of more than 4700 subjects (Ekelund, Lichtermann, Jarvelin, & Peltonen, 1999). Both of these studies suggest that our original observation of an association between the DRD4 48 base pair repeat region and TPQ NS may have been due to linkage disequilibrium between this polymorphism and another polymorphism within the DRD4 gene (the promoter region?) or in a neighboring gene. Finally, the first genome-wide scan in personality genetics in a group of alcoholic families (Cloninger *et al.*, 1998), which reported a linkage between anxiety-related traits and a region on chr 8p, also observed a linkage between NS and the region on chr 11 where the DRD4 gene is located (Cloninger, 2000, personal communication). This finding is currently being replicated in the second panel from this cohort.

Mutant mice lacking the DRD4 receptor also provide some evidence for a Novelty Seeking connection and this gene. In addition to increased sensitivity to ethanol (Rubinstein *et al.*, 1997), these mice also exhibit reduced behavioral response to novel stimuli (Dulawa, Grandy, Low, Paulus, & Geyer, 1999). The DRD4 receptor has also been analyzed in dog breeds and the long D allele was dominant in the endogenous Japanese breed (Shiba) compared to Golden retriever (Niimi, Inoue-Murayama, Murayama, Ito, & Iwasaki, 1999). It is interesting that the Shiba is considered an aggressive species and, as discussed above, some kinds of aggression are associated with Sensation/Novelty Seeking personality traits in humans.

Several reviews examine the evidence that the D4DR receptor plays a role in the partial determination of Novelty Seeking (Baron, 1998; Ebstein & Belmaker, 1997; Ebstein, Benjamin, & Belmaker, 2000; Malhotra & Goldman, 2000; Paterson, Sunohara, & Kennedy, 1999).

How many Genes Contribute to Novelty Seeking?

The effect size of the DRD4 exon III polymorphism on Novelty Seeking is modest, explaining only a small percentage (~5%) of the variance between individuals. Since genes account for perhaps 50% of the variance, and since it is likely that no major gene effect is present, it is likely that as many as 10 or more genes are contributing to this complex behavioral trait. These considerations prompted us to examine additional common polymorphisms for a role in Novelty Seeking and other personality dimensions. We first examined the 5-HT2C serotonin receptor polymorphism and its interaction with two dopamine receptor polymorphisms (D4DR exon III and D3DR gly/ser) and found an effect on TPQ Reward Dependent behavior (Ebstein, Segman *et al.*, 1997). Our study showed that when present in the same individual, the 5-HT2C and dopamine receptor polymorphisms account for 30% of the observed variance for persistence (RD2) and 13% of the variance for reward dependence scores (RD134). Similar findings were reported by a German group in a middle European population (Kuhn *et al.*, 1999).

Noble *et al.* (1998) showed that subjects with the DRD4 seven repeat allele had a significantly higher Novelty Seeking score than those without this allele. However, the greatest difference in Novelty Seeking score was found when subjects having all three minor DRD2 alleles and the DRD4 seven repeat allele were contrasted to those without any of these alleles. The combined DRD2 and DRD4 polymorphisms contribute more markedly to this behavior than when these two gene polymorphisms are individually considered.

Two other common polymorphisms of particular interest are the serotonin transporter promoter region (5-HTTLPR) and catechol-O-methyltransferase (COMT).

5-HTTLPR

Following our report of an association between DRD4 and NS, Lesch and his colleagues showed an association between a newly-discovered 44 bp deletion in the promoter region of the serotonin transporter promoter region (5-HTTLPR) affecting transcription of the gene (Heils *et al.*, 1996) and associated with neuroticism, or Harm Avoidance personality traits (Lesch *et al.*, 1996). Similar to investigations of the DRD4 receptor and Novelty Seeking, some subsequent reports replicated the initial association (Greenberg *et al.*, 2000; Katsuragi *et al.*, 1999; Murakami *et al.*, 1999; Osher, Hamer, & Benjamin, 2000; Ricketts *et al.*, 1998), whereas two yielded ambiguous results (Gelernter, Kranzler, Coccaro, Siever, & New, 1998; Mazzanti *et al.*, 1998). Several studies failed to observe any association between anxiety-related traits and the 5-HTTLPR polymorphism (Ball *et al.*, 1997; Deary *et al.*, 1999; Ebstein, Gritsenko *et al.*, 1997; Flory *et al.*, 1999; Gustavsson *et al.*, 1999; Hamilton *et al.*, 1999; Jorm *et al.*, 1998; Kumakiri *et al.*, 1999).

Serotonin neurotransmission has been postulated to play an important role in aggressive, violent and impulsive behavior in man (Berman, Tracy, & Coccaro, 1997; Kavoussi, Armstead, & Coccaro, 1997), and some studies have shown that platelet and central measures of serotonergic function are altered in aggressive individuals (Coccaro, Kavoussi, Sheline, Berman, & Csernansky, 1997; Coccaro, Kavoussi, Trestman *et al.*, 1997; for a more extensive discussion, see Chapter 9). The short 5-HTTLPR promoter polymorphism was also reported to be associated with an increased risk for early onset alcoholism, and with antisocial personality disorder and impulsive, habitually violent behavior (Hallikainen *et al.*, 1999). Consistent with an association between the promoter region polymorphism and antisocial personality disorder, two investigations (Seeger, Schloss, & Schmidt, 2001) including one from our group (Manor *et al.*, 2001) show an increased prevalence of the long/long 5-HTTLPR genotype in Attention Deficit Hyperactive Disorder (ADHD). ADHD is a childhood disorder that in some children and adolescents, especially those with comorbid conduct disorder, confers risk for adult violence and impulsive behaviors.

COMT

Since catechol-O-methyltransferase (COMT), along with monoamine oxidase (MAO), is one of the two enzymes that account for the initial steps in the degradation of catecholamines (Boulton & Eisenhofer, 1998), this gene is *ipso facto* a worthwhile candidate for influencing behavior in man. The physiological substrates of COMT include dopamine, norepinephrine and adrenaline, but in contrast to MAO, serotonin is not a substrate for this enzyme (Mannisto, 1998). A common COMT polymorphism coding for a thermolabile low activity enzyme has been recognized for two decades (Lotta *et al.*, 1995; Weinshilboum & Dunnette, 1981). A single amino acid substitution (val_{108}/met) in exon IV accounts for differences in thermolability: *val/val* homozygotes display high, *val/met* intermediate and *met/met* confers 4-5 times lower COMT activity (Lachman *et al.*, 1996).

Recently, homozygosity of a low enzyme-activity *met* variant of the COMT gene was reported to be associated with aggressive behavior in a small group of schizophrenic patients (Lachman, Nolan, Mohr, Saito, & Volavka, 1998; Strous, Bark, Parsia, Volavka, & Lachman, 1997). We have now confirmed this finding in a group of homicidal schizophrenic patients (Kotler *et al.*, 1999). Significant differences were observed in the distribution of COMT genotypes between homicidal inmates and schizophrenic patients at low risk for violent behavior. As previously reported, no difference in the distribution of COMT genotypes was observed between low-risk schizophrenic patients and controls. No association was observed between homicidal behavior and either the dopamine D4 exon III repeat length polymorphism or the 5-HTTLPR polymorphism. Since serotonin is not a substrate for COMT, these results suggest the involvement of non-serotonergic pathways such as dopamine and norepinephrine in such behaviors.

It is also worth noting that the high enzyme activity COMT *val* allele was initially reported to be in excess in polysubstance abusers from an American population using a case-control design (Vandenbergh, Rodriguez, Miller, Uhl, & Lachman, 1997). We recently replicated this finding in a group of Israeli heroin addicts using a family based design (Horowitz *et al.*, 2000).

Most intriguingly, a recent study (Egan *et al.*, 2001) of COMT demonstrated that this genotype was related in allele dosage fashion to performance on the Wisconsin Card Sorting Test of executive cognition, and explained 4% of the variance (p = 0.001) in frequency of perseverative errors. Consistent with other evidence that dopamine enhances prefrontal neuronal function, the load of the low-activity *met* allele predicted enhanced cognitive performance. We have examined the COMT polymorphism for its role in contributing to hypnotizability, a trait considered relevant to executive function (Crawford, Brown, & Moon, 1993), and we also found an association with the COMT genotype (Lichtenberg, Bachner-Melman, Gritsenko, & Ebstein, 2000). These results provide a genetic and neurochemical basis for interpreting a recent meta-analysis that showed that overall, antisocial groups performed 0.62 standard deviations worse on Executive Function tests than comparison groups; this effect size is in the medium to large range (Morgan & Lilienfeld, 2000).

DRD4 x COMT x 5-HTTLPR and Novelty Seeking

We examined the role of three functional polymorphisms: DRD4 exon III, serotonin transporter promoter region (5-HTTLPR) and catechol O-methyltransferase (COMT) polymorphisms, for association with TPQ personality factors in 455 subjects (Benjamin *et al.*, 2000). Our results (Benjamin *et al.*, 2000) showed that in the absence of the short 5-HTTLPR allele and in the presence of the high enzyme activity COMT *val/val* genotype, NS scores are higher in the presence of the DRD4 seven-repeat allele. The effect of these three polymorphisms on NS was also examined using a within-families design. Siblings who shared identical genotype groups for all three polymorphisms (COMT, DRD4 and 5-HTTLPR) had significantly correlated NS scores, whereas sibs with dissimilar genotypes in at least one polymorphism showed no significant correlation for NS scores. Similar interactions were also observed between these three polymorphisms and Novelty Seeking when 150 independently recruited and non-related subjects were analyzed.

THE BROAD DRD4 PHENOTYPE

So far we have discussed the role of the DRD4 exon III receptor polymorphism and its relevancy to the personality trait of

Novelty Seeking. However, this polymorphism has also been investigated in related psychiatric disorders, particularly attention deficit hyperactivity disorder (ADHD) and substance abuse. In both ADHD and drug-seeking behavior, dopaminergic neurotransmission is important and it is not surprising, therefore, that some studies discussed below implicate DRD4 in these related phenotypes.

ADHD

Soon after the first reports appeared linking the DRD4 exon III repeat allele with Novelty Seeking (Benjamin *et al.*, 1996; Ebstein *et al.*, 1996), this same polymorphism was shown to be associated with a behavioral disorder in children (LaHoste *et al.*, 1996), attention deficit hyperactivity disorder (ADHD)—a syndrome characterized in most patients with an impulsive phenotype related to adult novelty or sensation seeking personality dimensions. Indeed, adult patients with ADHD score high on TPQ Novelty Seeking (Downey, Stelson, Pomerleau, & Giordani, 1997). The significant but weak association between ADHD and the DRD4 receptor has been demonstrated by a recent meta-analysis, including both case-control and family studies (Faraone, Doyle, Mick, & Biederman, 2001).

Interestingly, common comorbid conditions of childhood and adolescent ADHD that persist into adulthood are conduct disorder (Faraone, Biederman, & Monuteaux, 2000) and substance abuse (Biederman *et al.*, 1995, 1997).

Addiction

Following our first report of an association between a particular personality trait and the DRD4 gene, we hypothesized that genotyping a group of individuals, such as substance abusers, sharing an extreme phenotype relevant to Novelty Seeking, would be a good approach towards validating our initial findings, first demonstrated with self-report questionnaires. We used a case-control design, comparing 141 heroin addicts to 110 control subjects, and found that the 7-repeat allele is significantly over-represented in the opioid-dependent cohort and confers a relative risk (i.e., odds ratio) of 2.46 (Kotler *et al.*, 1997; Mel *et al.*, 1998). To the best of our knowledge, our study was the first to show an association between a specific genetic polymorphism and opioid addiction. Our finding has been independently verified: in a group of Chinese (Li *et al.*, 1997) as well as in a group of American heroin addicts (Vandenbergh *et al.*, 2000). However, at least one other investigation from middle Europe failed to replicate this finding (Franke *et al.*, 2000). Similarly, an excess of the long forms of the DRD4 receptor in alcoholics (Muramatsu, Higuchi, Murayama, Matsushita, & Hayashida, 1996), especially in patients carrying the ALDH2(2) allele, was shown to be protective for alcoholism (Goldman, 1995). These results are consistent with the behavioral phenotype of a DRD4 knockout mouse that was characterized as supersensitive to ethanol, cocaine and methamphetamine (Rubinstein *et al.*, 1997).

However, in three studies with Caucasian populations (Geijer *et al.*, 1997; Malhotra *et al.*, 1996; Sander *et al.*, 1997) and one study in Taiwan (Chang, Ko, Lu, Pakstis, & Kidd, 1997), no evidence for an association between DRD4 and alcoholism was found.

Shields and his colleagues (Shields *et al.*, 1998) examined the DRD4 genotype for an association with smoking, in subjects following a single minimal-contact session of smoking cessation counseling, who were then were followed for up to 1 year. African-Americans (n = 72) who had at least one DRD4 long allele had a higher risk of smoking (odds ratio = 7.7; p = 0.006), waited less time for the first cigarette in the morning (p = 0.03), and started smoking younger (p = 0.09) than homozygote genotypes for the short-short alleles. After smoking cessation counseling, none of the African-American smokers with a long DRD4 allele was abstinent at 2 months, compared with 35% of the smokers homozygote for the short genotype (p = 0.02). The analysis of Caucasians did not suggest a similar smoking risk for the DRD4 genotypes.

Genes and Temperament in Early Human Development

We have also examined the very earliest manifestations of personality and the role of genes in the determination of temperament in neonates, infants and early childhood (Auerbach *et al.*, 1999; Auerbach, Benjamin, Faroy, Geller, & Ebstein, 2001; Ebstein *et al.*, 1998). We first examined the role of both the D4DR and 5-HTTLPR in a group of 2 week old neonates, using Brazelton's Neonate Behavioral Assessment Scale (NBAS) (Ebstein *et al.*, 1998). Although we found a significant effect, using both multivariate and univariate tests, for the D4DR exon III repeat polymorphism on the core temperament scales of the NBAS, no direct effect of the 5-HTTLPR polymorphism was observed. The effect of the homozygous short 5-HTTLPR genotype was to lower the orientation score for the group of neonates lacking the long form of D4DR. Orientation may be considered to be a behavior that stimulates interaction with the environment, and is predominant or constitutive in infancy. Orientation promotes active searching and acquaintance with the neonate's new environment. We have suggested that NBAS Orientation may be a very early precedent of adult Novelty Seeking.

These same infants were also evaluated at two months of age (Auerbach *et al.*, 1999). Infants with long D4DR alleles had significantly lower scores on Negative Emotionality and Distress to Limitations than infants with short D4DR alleles. In contrast, infants with the short homozygous 5-HTTLPR genotype had higher scores on Negative Emotionality and Distress to Limitations than infants with the L/S or L/L genotypes. The strongest effects occurred in those infants homozygous for the short 5-HTTLPR polymorphism who also were lacking long D4DR alleles. These infants showed most negative emotionality and most distress to daily situations—temperament traits that are perhaps the underpinning of adult neuroticism.

The dopamine D4 receptor (DRD4) exon III polymorphism has generated interest because of its association with ADHD, with an increased frequency of the seven-repeat allele being reported in children with ADHD (see discussion below). Deficits in sustained attention and information processing characterize ADHD, and individual differences in these functions are apparent from infancy. We found (Auerbach *et al.*, 2001) that in a structured play situation and on an information-processing task, 1-year-old infants with the DRD4 exon III long 7 repeat allele showed less sustained attention and novelty preference than do infants without this allele. There was also a significant interaction between DRD4 and the serotonin transporter promoter (5-HTTLPR) gene on a measure of sustained attention. Our results provide evidence for a possible developmental link between DRD4 and ADHD via early-sustained attention and information processing.

Following our first studies of the role of the DRD4 receptor in early infant development, Lakatos *et al.* (2000) also examined the role of the 7-repeat allele in disorganized attachment behavior. About 15% of one-year-old infants in non-clinical, low-risk, and up to 80% in high-risk (e.g., maltreated) populations show extensive disorganized attachment behavior in the Strange Situation Test. It has also been reported that disorganization of early attachment is a major risk factor for the development of childhood behavior problems. The collapse of organized attachment strategy has been explained primarily by inappropriate care giving, but recently the contribution of child factors such as neurological impairments and neonatal behavioral organization has also been suggested. Attachment behavior of 90 infants was tested in the Strange Situation and they were independently genotyped for the number of the 48-bp repeats. The 7-repeat allele was represented with a significantly higher frequency in infants classified as disorganized, compared to non-disorganized infants. The estimated relative risk for disorganized attachment among children carrying the 7-repeat allele was 4.15.

De Luca *et al.* (2001) examined the relationship between the DRD4 receptor and infants' behavior, measured with the Italian version of the Early and Revised Infancy Temperament Questionnaires in 122 Italian neonates at 1 and 5 months of life. Two-way (genotype x age) analysis of variance revealed a significant correlation with the temperamental subscale of adaptability. At 1 month of life (early assessment), infants with long DRD4 alleles presented significantly low scores in comparison with the high scores of infants with short alleles. These differences were not detected at 5 months of life (late assessment), which the authors interpret as denoting a strong environmental effect at this age on the genetic background. Table 1 summarizes 5 molecular-genetic studies of early human temperament.

Interpreting these five studies is confounded by the several instruments that were employed in assessing early temperament across three distinct ethnic groups (Israeli, Italian and Hungarian). Moreover, the first year of life is characterized by radical and rapid changes in the infant's behavior, further obscuring associations between genes and phenotype. Nevertheless, it appears that at least by 12 months of age the presence of the long forms of the DRD4 allele leads to behaviors that can be construed as precedents to maladaptive traits in later developmental times.

Common Genetic Polymorphisms and Arousal Theory

We have reviewed evidence that during development from infancy to adulthood it is possible to discern the modest effect of the DRD4 and other common polymorphisms on several correlated behaviors. Surprisingly, five studies have shown that it is possible to observe the influence on behavior of the long DRD4 alleles in the first year of development. In childhood and adolescence, an excess of the long DRD4 alleles characterizes the group of subjects diagnosed with ADHD. In adulthood, some subjects with high scores on the Novelty or Sensation Seeking trait measured by self-report personality questionnaires, and some individuals characterized by drug-seeking behavior, also show a higher frequency of the long DRD4 alleles.

Table 1. Molecular-genetic studies of early human temperament.

Age	Test	Results	Reference
2 Weeks	Brazelton's NBAS	+DRD4 7 allele increases scores on core NBAS clusters; interaction with 5 HTTLPR	Ebstein *et al.* (1998)
2 months	Infant Behavioral Questionnaire	+DRD4 7 allele << scores on Negative Emotionality and Distress to Limitations; (s/s) 5-HTTLPR >> scores on Negative Emotionality and Distress to Limitations; +++ effects with the s/s 5-HTTLPR polymorphism also lacking long D4DR alleles	Auerbach *et al.* (1999)
12 months	Structured play situation and information-processing task	+DRD4 7 allele << sustained attention and novelty preference; significant interaction between DRD4 & 5-HTTLPR on a measure of sustained attention.	Auerbach *et al.* (2001)
12 months	Strange Situation Test	+DRD4 7 repeat allele over represented in infants with disorganized attachment	Lakatos *et al.* (2000)
1 & 5 months	Early and Revised Infancy Temperament Questionnaires	At one month +DRD4 7 allele << adaptability; no effect of genotype at 5 months	De Luca *et al.* (2001)

Note. << *= less;* + *= presence of the allele;* >> *= increases scores;* +++ *= greater effect*

How are we best to understand these effects of the DRD4 gene on behavior? Zuckerman (1994) cogently discusses the notion that Sensation or Novelty Seeking is a primary drive in both animals and man, that has its expression in human personality, and furthermore is closely linked to brain reward mechanisms in which dopamine neurons play a critical role (Kalivas & Nakamura, 1999; Robbins & Everitt, 1996; Schultz, 1997; Wise, 1996; see also Chapter 2). Arousal theories propose that low intensities of stimulation tend to evoke approach reactions and high intensities of stimulation withdrawal reactions. Approach-Withdrawal has also been identified as a fundamental trait in human neonates (Brazelton, Tronick, Adamson, Als, & Wise, 1975) and may be the first manifestation of sensation seeking in humans. Stimuli of any variety (and not necessarily primary reward objects like sex or food) possess *motivation value*—a property not only of stimulus *intensity* but also of *arousal potential*. Arousal potential represents the capacity of stimuli to command attention, excite the nervous system or influence behavior. Arousal potential is particularly dependent on novelty, degree of change from preceding stimulation, suddenness of change, surprise, incongruity and uncertainty (cf. Zuckerman, 1994). The motive circuit in the brain heavily dependent on dopaminergic transmission provides the neuroanatomical and neurochemical basis for the reward value of arousing stimuli.

It has been suggested that there are stable, individual differences in optimal levels of stimulus arousal preference that serve as the foundation for basic human personality traits. For example, Eysenck proposed that extraverts require higher levels of stimulation compared to introverts, and since such a high level is not provided by routine environments they would tend to be sensation or novelty seekers (cf. Zuckerman, 1994). Introverts reach optimal levels of arousal at lower levels of stimulation than the population average. Differences between extraverts and introverts are presumed to be due to individual variations in background, non-stimulated or tonic levels of arousal (see also Chapter 1). *Such tonic levels of arousal may be determined at least partially by common genetic polymorphisms as discussed in this review.* One of these genes is the DRD4 receptor, but two other common polymorphisms that have been widely studied for their relevancy to human behavior, COMT and 5-HTTLPR, are also likely to contribute to the determination of personality and cognitive traits that predispose to violent behavior.

Relevant to our discussion is a considerable body of evidence that points to autonomic *underarousal* for individuals characterized by antisocial personality disorders (Raine, 1996; see Chapter 8 for a comprehensive review). Criminals have been shown to display a significantly lower resting heart rate, skin conductance activity, and more slow-frequency electroencephalographic activity than non-criminals (Raine, Venables, & Williams, 1990). Moreover, these differences were not mediated by social, demographic, and academic factors. Recent findings (Raine, Lencz, Bihrle, LaCasse, & Colletti, 2000) provide the first evidence for a structural brain deficit in APD (APD group showed an 11.0% reduction in prefrontal gray matter volume). This prefrontal structural deficit may underlie the low arousal, poor fear conditioning, lack of conscience, and decision-making deficits that have been found to characterize antisocial, psychopathic behavior. It is not unlikely, moreover, that these differences in autonomic arousal and brain anatomy have a partial genetic basis.

Complex Interactions Between Genes, Personality Traits, and Aggression

We have discussed the evidence that Novelty or Sensation Seeking is a prominent personality trait in many subject groups exhibiting psychopathic behaviors. We have shown that this personality trait is measurable using any one of several self-report questionnaires. Moreover, Novelty or Sensation Seeking is heritable—as much as genes explain 50% of the variance in this trait. We have also discussed recent findings, initially from this laboratory, that a common genetic polymorphism can account for some of the genetic variance in this trait. Moreover, we and others have demonstrated that more than one gene of small effect sizes are contributing to personality traits. We have also emphasized that Novelty or Sensation Seeking is part of the architecture of normal personality, and that many high scorers on this trait are productive and adaptive members of society.

Why, then, do only a minority of subjects exhibiting the trait of Novelty or Sensation Seeking progress to criminal and anti-social behaviors? The 7.5-15% of the population with Novelty or Sensation Seeking scores 1-2 standard deviations from the mean are mainly individuals expressing this temperament by mountain climbing, martial arts training, bungy jumping, flying airplanes and perhaps pursuing creative careers in the sciences and the arts. Studies by Cloninger suggest that the non-heritable elements of personality, defined by character, are important modulators of the more heritable temperament traits. As discussed in the next section, Cloninger has invented a comprehensive personality questionnaire that ascertains both temperament and character traits.

A note of caution is needed, however, regarding Cloninger's contention that the non-temperament traits in the TCI are less heritable than the temperament traits such as Novelty Seeking. For example, Comings and his colleagues (Comings, Gonzales, Saucier, Johnson, & MacMurray, 2000) have recently shown that TCI self-transcendence scores are correlated with DRD4 genotype.

Cloninger's Seven Factor Model: Interaction Between Heritable and non-Heritable Traits in the Architecture of Personality

Cloninger defines personality as the dynamic organization of the psychobiological system that modulates adaptation to experience (Cloninger, 1987). Two aspects or domains of personality have been recognized: temperament and character. Temperament may be viewed as the heritable, more autonomic domain of personality. Temperament defines the style or manner of an individual's behavior; it is emotionally based, and is relatively uninfluenced by socio-cultural learning. About 50% of the variance in temperament among individuals is heritable and stable from childhood through adulthood. In contrast, character is weakly heritable (10-15%), but is moderately influenced by sociocultural learning. Character involves differences in higher cognitive processing, such as concepts about self and relations to others. It encompasses traits such as self-directed behavior, empathic social cooperation and creative symbolic invention. The TCI personality questionnaire defines three character dimensions: self-directedness (responsible, purposeful, resourceful), cooperative (tender hearted, empathic, helpful) and self-transcendent (self-forgetful, transpersonal, spiritual, enlightened). In addition, the TCI defines the four temperament traits of Novelty Seeking (exploratory, curious, impulsive, extravagant), Harm Avoidance (worrying & pessimistic, fearful, shy, fatigable), Reward Dependence (sentimental, dedicated, dependent) and Persistence (industrious, hard-working, ambitious, perfectionist).

We concur with Cloninger that it is the character elements in personality architecture that are likely to modulate more heritable temperament elements, and partially explain the positive and negative consequences in individuals exhibiting particular temperaments such as

Novelty or Sensation Seeking. Moreover, we expect that the relation to genetic polymorphisms will be strongest for heritable temperament traits and weaker or non-significant for character dimensions. Likewise, we expect that 'positive' Sensation Seekers (e.g., athletes, army commandos) will score significantly higher than 'negative' Sensation Seekers (e.g., criminals, sexual offenders) on *character* traits such as self-directedness (responsible, purposeful, resourceful), cooperative (tender hearted, empathic, helpful) and self-transcendent (self-forgetful, transpersonal, spiritual, enlightened) scales.

Cloninger's character factors can be interpreted as the result of numerous environmental influences—parents, home, neighborhood, socio-economic status, school, religious education, etc. The character facets of personality are formed following a long, slow developmental process of social learning by the individual as he or she matures. One likely explanation for subjects with 'poor' character traits is that they are not provided with the appropriate sociocultural learning cues due to an impoverished environment (for a relevant discussion, see Chapter 5). Impoverished environment generally involves the nuclear family (poor parenting), and several adoption studies underscore this complex interacting relationship between genes and environment. Individuals at risk for psychopathic behavior and violence show a temperamental genetic diathesis for this behavior. Coupled with an environment providing the wrong sociocultural learning experience, this temperamental predisposition ensures the emergence of the violent phenotype. Cadoret examined the origins of conduct disorder and aggressiveness and showed that in the absence of a biological parent with ASPD there is no correlation between adoptee outcome (conduct disorder and adult ASPD) and adverse adoptive home environment (Cadoret, Yates, Troughton, Woodworth, & Stewart, 1995). In contrast, the adverse adoptive home environment-outcome relationship is significant only for adoptees with a biological parent with ASPD. ASPD is increased in the presence of predisposing genetic factors. Genes are powerful but they exert their effects in the context of a hierarchy of micro and macro environments ranging from cellular genetic milieu to the family and neighborhood school.

REFERENCES

Arai, Y., Hosokawa, T., Fukao, A., Izumi, Y., & Hisamichi, S. (1997). Smoking behaviour and personality: a population-based study in Japan. *Addiction*, **92**, 1023-1033.

Asghari, V., Sanyal, S., Buchwaldt, S., Paterson, A., Jovanovic, V., & Van Tol, H.H. (1995). Modulation of intracellular cyclic AMP levels by different human dopamine D4 receptor variants. *Journal of Neurochemistry*, **65**, 1157-1165.

Asghari, V., Schoots, O., van Kats, S., Ohara, K., Jovanovic, V., Guan, H.C., Bunzow, J.R., Petronis, A., & Van Tol, H.H. (1994). Dopamine D4 receptor repeat: analysis of different native and mutant forms of the human and rat genes. *Molecular Pharmacology*, **46**, 364-373.

Auerbach, J., Geller, V., Lezer, S., Shinwell, E., Belmaker, R.H., Levine, J., & Ebstein, R. (1999). Dopamine D4 receptor (D4DR) and serotonin transporter promoter (5-HTTLPR) polymorphisms in the determination of temperament in 2-month-old infants. *Molecular Psychiatry*, **4**, 369-373.

Auerbach, J.G., Benjamin, J., Faroy, M., Geller, V., & Ebstein, R. (2001). DRD4 related to infant attention and information processing: a developmental link to ADHD? *Psychiatric Genetics*, **11**, 31-35.

Ball, D., Hill, L., Freeman, B., Eley, T.C., Strelau, J., Riemann, R., Spinath, F.M., Angleitner, A., & Plomin, R. (1997). The serotonin transporter gene and peer-rated neuroticism. *Neuroreport*, **8**, 1301-1304.

Baron, M. (1998). Mapping genes for personality: Is the saga sagging? *Molecular Psychiatry*, **3**, 106-108.

Bau, C.H., Almeida, S., Costa, F.T., Garcia, C.E., Elias, E.P., Ponso, A.C., Spode, A., & Hutz, M.H. (2001). DRD4 and DAT1 as modifying genes in alcoholism: interaction with novelty seeking on level of alcohol consumption. *Molecular Psychiatry*, **6**, 7-9.

Bau, C.H., Roman, T., Almeida, S., & Hutz, M.H. (1999). Dopamine D4 receptor gene and personality dimensions in Brazilian male alcoholics. *Psychiatric Genetics*, **9**, 139-143.

Benjamin, J., Li, L., Patterson, C., Greenberg, B.D., Murphy, D.L., & Hamer, D.H. (1996). Population and familial association between the D4 dopamine receptor gene and measures of Novelty Seeking. *Nature Genetics*, 12, 81-84.

Benjamin, J., Osher, Y., Kotler, M., Gritsenko, I., Nemanov, L., Belmaker, R.H., & Ebstein, R.P. (2000). Association between tridimensional personality questionnaire (TPQ) traits and three functional polymorphisms: dopamine receptor D4 (DRD4), serotonin transporter promoter region (5-HTTLPR) and catechol O-methyltransferase (COMT). *Molecular Psychiatry*, **5**, 96-100.

Berman, M.E., Tracy, J.I., & Coccaro, E.F. (1997). The serotonin hypothesis of aggression revisited. *Clinical Psychology Review*, **17**, 651-665.

Biederman, J., Wilens, T., Mick, E., Faraone, S.V., Weber, W., Curtis, S., Thornell, A., Pfister, K., Jetton, J.G., & Soriano, J. (1997). Is ADHD a risk factor for psychoactive substance use disorders? Findings from a four-year prospective follow-up study. *Journal of the American Academy of Child Adolescence and Psychiatry*, **36**, 21-29.

Biederman, J., Wilens, T., Mick, E., Milberger, S., Spencer, T.J., & Faraone, S.V. (1995). Psychoactive substance use disorders in adults with attention deficit hyperactivity disorder (ADHD): effects of ADHD and psychiatric comorbidity. *American Journal of Psychiatry*, **152**, 1652-1658.

Blackburn, R. (1987). Two scales for the assessment for personality disorder in antisocial populations. *Personality and Individual Differences*, **8**, 81-93.
Blaszczynski, A., Steel, Z., & McConaghy, N. (1997). Impulsivity in pathological gambling: the antisocial impulsivist. *Addiction*, **92**, 75-87.
Bouchard, T.J., Jr. (1994). Genes, environment, and personality. *Science*, **264**, 1700-1701.
Boulton, A.A., & Eisenhofer, G. (1998). Catecholamine metabolism. From molecular understanding to clinical diagnosis and treatment. Overview. *Advances in Pharmacology*, **42**, 273-292.
Brazelton, T.B., Tronick, E., Adamson, L., Als, H., & Wise, S. (1975). Early mother-infant reciprocity. *Ciba Foundation Symposium*, 33, 137-154.
Cadoret, R.J., Yates, W.R., Troughton, E., Woodworth, G., & Stewart, M.A. (1995). Genetic-environmental interaction in the genesis of aggressivity and conduct disorders. *Archives of General Psychiatry*, **52**, 916-924.
Carton, S., Jouvent, R., & Widlocher, D. (1994). Sensation seeking, nicotine dependence, and smoking motivation in female and male smokers. *Addictive Behavior*, **19**, 219-227.
Chang, F.M., Ko, H.C., Lu, R.B., Pakstis, A.J., & Kidd, K.K. (1997). The dopamine D4 receptor gene (DRD4) is not associated with alcoholism in three Taiwanese populations: six polymorphisms tested separately and as haplotypes. *Biological Psychiatry*, **41**, 394-405.
Cherek, D.R., Moeller, F.G., Dougherty, D.M., & Rhoades, H. (1997). Studies of violent and nonviolent male parolees: II. Laboratory and psychometric measurements of impulsivity. *Biological Psychiatry*, **41**, 523-529.
Cloninger, C.R. (1986). A unified biosocial theory of personality and its role in the development of anxiety states. *Psychiatric Development*, **4**, 167-226.
Cloninger, C.R. (1987). A systematic method for clinical description and classification of personality variants. A proposal. *Archives of General Psychiatry*, **44**, 573-588.
Cloninger, C.R., Van Eerdewegh, P., Goate, A., Edenberg, H.J., Blangero, J., Hesselbrock, V., Reich, T., Nurnberger, J., Jr., Schuckit, M., Porjesz, B., Crowe, R., Rice, J.P., Foroud, T., Przybeck, T.R., Almasy, L., Bucholz, K., Wu, W., Shears, S., Carr, K., Crose, C., Willig, C., Zhao, J., Tischfield, J.A., Li, T.K., Conneally, P.M., *et al.* (1998). Anxiety proneness linked to epistatic loci in genome scan of human personality traits. *American Journal of Medical Genetics*, **81**, 313-317.
Coccaro, E.F., Kavoussi, R.J., Sheline, Y.I., Berman, M.E., & Csernansky, J.G. (1997). Impulsive aggression in personality disorder correlates with platelet 5-HT2A receptor binding. *Neuropsychopharmacology*, **16**, 211-216.
Coccaro, E.F., Kavoussi, R.J., Trestman, R.L., Gabriel, S.M., Cooper, T.B., & Siever, L.J. (1997). Serotonin function in human subjects: intercorrelations among central 5- HT indices and aggressiveness. *Psychiatry Research*, **73**, 1-14.
Comings, D.E., Gonzales, N., Saucier, G., Johnson, J.P., & MacMurray, J.P. (2000). The DRD4 gene and the spiritual transcendence scale of the character temperament index. *Psychiatric Genetics*, **10**, 185-189.
Crawford, H.J., Brown, A.M., & Moon, C.E. (1993). Sustained attentional and disattentional abilities: Differences between low and highly hypnotizable persons. *Journal of Abnormal Psychology*, **102**, 534-543.
Cruz, C., Camarena, B., Mejia, J.M., Paez, F., Eroza, V., Ramon De La Fuente, J., Kershenobich, D., & Nicolini, H. (1995). The dopamine D2 receptor gene TaqI A1 polymorphism and alcoholism in a Mexican population. *Archives of Medical Research*, **26**, 421-426.

De Luca, A., Rizzardi, M., Torrente, I., Alessandroni, R., Salvioli, G.P., Filograsso, N., Dallapiccola, B., & Novelli, G. (2001). Dopamine D4 receptor (DRD4) polymorphism and adaptability trait during infancy: a longitudinal study in 1- to 5-month-old neonates. *Neurogenetics,* **3**, 79-82.

Deary, I.J., Battersby, S., Whiteman, M.C., Connor, J.M., Fowkes, F.G., & Harmar, A. (1999). Neuroticism and polymorphisms in the serotonin transporter gene. *Psychological Medicine,* **29**, 735-739.

Downey, K.K., Stelson, F.W., Pomerleau, O.F., & Giordani, B. (1997). Adult attention deficit hyperactivity disorder: psychological test profiles in a clinical population. *Journal of Nervous and Mental Disease,* **185**, 32-38.

Dulawa, S.C., Grandy, D.K., Low, M.J., Paulus, M.P., & Geyer, M.A. (1999). Dopamine D4 receptor-knock-out mice exhibit reduced exploration of novel stimuli. *Journal of Neuroscience,* **19**, 9550-9556.

Ebstein, R.P., & Belmaker, R.H. (1997). Saga of an adventure gene: novelty seeking, substance abuse and the dopamine D4 receptor (D4DR) exon III repeat polymorphism. *Molecular Psychiatry,* **2**, 381-384.

Ebstein, R.P., Benjamin, J., & Belmaker, R.H. (2000). Personality and polymorphisms of genes involved in aminergic neurotransmission. *European Journal of Pharmacology,* **410**, 205-214.

Ebstein, R.P., Gritsenko, I., Nemanov, L., Frisch, A., Osher, Y., & Belmaker, R.H. (1997). No association between the serotonin transporter gene regulatory region polymorphism and the Tridimensional Personality Questionnaire (TPQ) temperament of harm avoidance. *Molecular Psychiatry,* **2**, 224-226.

Ebstein, R.P., Levine, J., Geller, V., Auerbach, J., Gritsenko, I., & Belmaker, R.H. (1998). Dopamine D4 receptor and serotonin transporter promoter in the determination of neonatal temperament. *Molecular Psychiatry,* **3**, 238-246.

Ebstein, R.P., Novick, O., Umansky, R., Priel, B., Osher, Y., Blaine, D., Bennett, E.R., Nemanov, L., Katz, M., & Belmaker, R.H. (1996). Dopamine D4 receptor (D4DR) exon III polymorphism associated with the human personality trait of Novelty Seeking. *Nature Genetics,* **12**, 78-80.

Ebstein, R.P., Segman, R., Benjamin, J., Osher, Y., Nemanov, L., & Belmaker, R.H. (1997). 5-HT2C (HTR2C) serotonin receptor gene polymorphism associated with the human personality trait of reward dependence: interaction with dopamine D4 receptor (D4DR) and dopamine D3 receptor (D3DR) polymorphisms. *American Journal of Medical Genetics,* **74**, 65-72.

Egan, M.F., Goldberg, T.E., Kolachana, B.S., Callicott, J.H., Mazzanti, C.M., Straub, R.E., Goldman, D., & Weinberger, D.R. (2001). Effect of COMT Val108/158 Met genotype on frontal lobe function and risk for schizophrenia. *Proceedings of the National Academy of Sciences of the USA,* **98**, 6917-6922.

Ekelund, J., Lichtermann, D., Jarvelin, M.R., & Peltonen, L. (1999). Association between Novelty Seeking and the Type 4 dopamine receptor gene in a large Finnish cohort sample. *American Journal of Psychiatry,* **156**, 1453-1455.

Eysenck, H.J. (1998). *Dimensions of personality.* New Brunswick, NJ: Transaction Publishers.

Faraone, S.V., Biederman, J., & Monuteaux, M.C. (2000). Toward guidelines for pedigree selection in genetic studies of attention deficit hyperactivity disorder. *Genetic Epidemiology,* **18**, 1-16.

Faraone, S.V., Doyle, A.E., Mick, E., & Biederman, J. (2001). Meta-analysis of the association between the 7-repeat allele of the dopamine D(4) receptor gene and attention deficit hyperactivity disorder. *American Journal of Psychiatry,* **158**, 1052-1057.

Ficher, I.V., Zuckerman, M., & Neeb, M. (1981). Marital compatibility in sensation seeking trait as a factor in marital adjustment. *Journal of Sex and Marital Therapy*, **7**, 60-69.
Floderus-Myrhed, B., Pedersen, N., & Rasmuson, I. (1980). Assessment of heritability for personality, based on a short-form of the Eysenck Personality Inventory: a study of 12,898 twin pairs. *Behavior Genetics*, **10**, 153-162.
Flory, J.D., Manuck, S.B., Ferrell, R.E., Dent, K.M., Peters, D.G., & Muldoon, M.F. (1999). Neuroticism is not associated with the serotonin transporter (5-HTTLPR) polymorphism. *Molecular Psychiatry*, **4**, 93-96.
Franke, P., Nothen, M.M., Wang, T., Knapp, M., Lichtermann, D., Neidt, H., Sander, T., Propping, P., & Maier, W. (2000). DRD4 exon III VNTR polymorphism-susceptibility factor for heroin dependence? Results of a case-control and a family-based association approach. *Molecular Psychiatry*, **5**, 101-104.
Gebhardt, C., Leisch, F., Schussler, P., Fuchs, K., Stompe, T., Sieghart, W., Hornik, K., Kasper, S., & Aschauer, H.N. (2000). Non-association of dopamine D4 and D2 receptor genes with personality in healthy individuals. *Psychiatric Genetics*, **10**, 131-137.
Geijer, T., Jonsson, E., Neiman, J., Persson, M.L., Brene, S., Gyllander, A., Sedvall, G., Rydberg, U., Wasserman, D., & Terenius, L. (1997). Tyrosine hydroxylase and dopamine D4 receptor allelic distribution in Scandinavian chronic alcoholics. *Alcoholism: Clinical and Experimental Research*, **21**, 35-39.
Gelernter, J., Kranzler, H., Coccaro, E., Siever, L., New, A., & Mulgrew, C.L. (1997). D4 dopamine-receptor (DRD4) alleles and novelty seeking in substance-dependent, personality-disorder, and control subjects. *American Journal of Human Genetics*, **61**, 1144-1152.
Gelernter, J., Kranzler, H., Coccaro, E.F., Siever, L.J., & New, A.S. (1998). Serotonin transporter protein gene polymorphism and personality measures in African American and European American subjects. *American Journal of Psychiatry*, **155**, 1332-1338.
Goldman, D. (1995). Candidate genes in alcoholism. *Clinical Neuroscience*, **3**, 174-181.
Greenberg, B.D., Li, Q., Lucas, F.R., Hu, S., Sirota, L.A., Benjamin, J., Lesch, K.P., Hamer, D., & Murphy, D.L. (2000). Association between the serotonin transporter promoter polymorphism and personality traits in a primarily female population sample. *American Journal of Medical Genetics*, **96**, 202-216.
Grosz, D.E., Lipschitz, D.S., Eldar, S., Finkelstein, G., Blackwood, N., Gerbino-Rosen, G., Faedda, G.L., & Plutchik, R. (1994). Correlates of violence risk in hospitalized adolescents. *Comprehensive Psychiatry*, **35**, 296-300.
Gustavsson, J.P., Nothen, M.M., Jonsson, E.G., Neidt, H., Forslund, K., Rylander, G., Mattila-Evenden, M., Sedvall, G.C., Propping, P., & Asberg, M. (1999). No association between serotonin transporter gene polymorphisms and personality traits. *American Journal of Medical Genetics*, **88**, 430-436.
Hallikainen, T., Saito, T., Lachman, H.M., Volavka, J., Pohjalainen, T., Ryynanen, O.P., Kauhanen, J., Syvalahti, E., Hietala, J., & Tiihonen, J. (1999). Association between low activity serotonin transporter promoter genotype and early onset alcoholism with habitual impulsive violent behavior. *Molecular Psychiatry*, **4**, 385-388.
Hamilton, S.P., Heiman, G.A., Haghighi, F., Mick, S., Klein, D.F., Hodge, S.E., Weissman, M.M., Fyer, A.J., & Knowles, J.A. (1999). Lack of genetic linkage or association between a functional serotonin transporter polymorphism and panic disorder. *Psychiatric Genetics*, **9**, 1-6.

Hare, R.D., & McPherson, L.M. (1984). Violent and aggressive behavior by criminal psychopaths. *International Journal of Law and Psychiatry*, **7**, 35-50.

Heath, A.C., Cloninger, C.R., & Martin, N.G. (1994). Testing a model for the genetic structure of personality: a comparison of the personality systems of Cloninger and Eysenck. *Journal of Personality and Social Psychology*, **66**, 762-775.

Heils, A., Teufel, A., Petri, S., Stober, G., Riederer, P., Bengel, D., & Lesch, K.P. (1996). Allelic variation of human serotonin transporter gene expression. *Journal of Neurochemistry*, **66**, 2621-2624.

Herbst, J.H., Zonderman, A.B., McCrae, R.R., & Costa, P.T., Jr. (2000). Do the dimensions of the temperament and character inventory map a simple genetic architecture? Evidence from molecular genetics and factor analysis. *American Journal of Psychiatry*, **157**, 1285-1290.

Horowitz, R., Kotler, M., Shufman, E., Aharoni, S., Kremer, I., Cohen, H., & Ebstein, R.P. (2000). Confirmation of an excess of the high enzyme activity COMT val allele in heroin addicts in a family-based haplotype relative risk study. *American Journal of Medical Genetics*, **96**, 599-603.

Howarth, E. (1986). What does Eysenck's psychoticism scale really measure? *British Journal of Psychology*, **77**, 223-227.

Hur, Y.M., & Bouchard, T.J., Jr. (1997). The genetic correlation between impulsivity and sensation seeking traits. *Behavior Genetics*, **27**, 455-463.

Hymbaugh, K., & Garrett, J. (1974). Sensation seeking among skydivers. *Perceptual and Motor Skills*, **38**, 118.

Jonah, B.A. (1997). Sensation seeking and risky driving: a review and synthesis of the literature. *Accident Analysis and Prevention*, **29**, 651-665.

Jonsson, E.G., Nothen, M.M., Gustavsson, J.P., Neidt, H., Forslund, K., Mattila-Evenden, M., Rylander, G., Propping, P., & Asberg, M. (1998). Lack of association between dopamine D4 receptor gene and personality traits. *Psychological Medicine*, **28**, 985-989.

Jorm, A.F., Henderson, A.S., Jacomb, P.A., Christensen, H., Korten, A.E., Rodgers, B., Tan, X., & Easteal, S. (1998). An association study of a functional polymorphism of the serotonin transporter gene with personality and psychiatric symptoms. *Molecular Psychiatry*, **3**, 449-451.

Kalivas, P.W., & Nakamura, M. (1999). Neural systems for behavioral activation and reward. *Current Opinion in Neurobiology*, **9**, 223-227.

Kaplan, H.I., Sadock, B.J., & Grebb, J.A. (1994). *Synopsis of psychiatry* (7th ed.). Baltimore: Williams and Wilkins.

Katsuragi, S., Kunugi, H., Sano, A., Tsutsumi, T., Isogawa, K., Nanko, S., & Akiyoshi, J. (1999). Association between serotonin transporter gene polymorphism and anxiety-related traits. *Biological Psychiatry*, **45**, 368-370.

Kavoussi, R., Armstead, P., & Coccaro, E. (1997). The neurobiology of impulsive aggression. *Psychiatric Clinics of North America*, **20**, 395-403.

Kazmi, M.A., Snyder, L.A., Cypess, A.M., Graber, S.G., & Sakmar, T.P. (2000). Selective reconstitution of human D4 dopamine receptor variants with Gi alpha subtypes. *Biochemistry*, **39**, 3734-3744.

Klinteberg, B., Hallman, J., Oreland, L., Wirsen, A., Levander, S.E., & Schalling, D. (1992). Exploring the connections between platelet monoamine oxidase activity and behavior. II. Impulsive personality without neuropsychological signs of disinhibition in air force pilot recruits. *Neuropsychobiology*, **26**, 136-145.

Koopmans, J.R., Boomsma, D.I., Heath, A.C., & van Doornen, L.J. (1995). A multivariate genetic analysis of sensation seeking. *Behavior Genetics*, **25**, 349-356.

Kotler, M., Barak, P., Cohen, H., Averbuch, I.E., Grinshpoon, A., Gritsenko, I., Nemanov, L., & Ebstein, R.P. (1999). Homicidal behavior in schizophrenia associated with a genetic polymorphism determining low catechol O-methyltransferase (COMT) activity. *American Journal of Medical Genetics*, **88**, 628-633.

Kotler, M., Cohen, H., Segman, R., Gritsenko, I., Nemanov, L., Lerer, B., Kramer, I., Zer-Zion, M., Kletz, I., & Ebstein, R.P. (1997). Excess dopamine D4 receptor (D4DR) exon III seven repeat allele in opioid-dependent subjects. *Molecular Psychiatry*, **2**, 251-254.

Kuhn, K.U., Meyer, K., Nothen, M.M., Gansicke, M., Papassotiropoulos, A., & Maier, W. (1999). Allelic variants of dopamine receptor D4 (DRD4) and serotonin receptor 5HT2c (HTR2c) and temperament factors: replication tests. *American Journal of Medical Genetics*, **88**, 168-172.

Kumakiri, C., Kodama, K., Shimizu, E., Yamanouchi, N., Okada, S., Noda, S., Okamoto, H., Sato, T., & Shirasawa, H. (1999). Study of the association between the serotonin transporter gene regulatory region polymorphism and personality traits in a Japanese population. *Neuroscience Letters*, **263**, 205-207.

La Grange, L., Jones, T.D., Erb, L., & Reyes, E. (1995). Alcohol consumption: biochemical and personality correlates in a college student population. *Addictive Behavior*, **20**, 93-103.

Lachman, H.M., Nolan, K.A., Mohr, P., Saito, T., & Volavka, J. (1998). Association between catechol O-methyltransferase genotype and violence in schizophrenia and schizoaffective disorder. *American Journal of Psychiatry*, **155**, 835-837.

Lachman, H.M., Papolos, D.F., Saito, T., Yu, Y.M., Szumlanski, C.L., & Weinshilboum, R.M. (1996). Human catechol-O-methyltransferase pharmacogenetics: description of a functional polymorphism and its potential application to neuropsychiatric disorders. *Pharmacogenetics*, **6**, 243-250.

LaHoste, G.J., Swanson, J.M., Wigal, S.B., Glabe, C., Wigal, T., King, N., & Kennedy, J.L. (1996). Dopamine D4 receptor gene polymorphism is associated with attention deficit hyperactivity disorder. *Molecular Psychiatry*, **1**, 121-124.

Lakatos, K., Toth, I., Nemoda, Z., Ney, K., Sasvari-Szekely, M., & Gervai, J. (2000). Dopamine D4 receptor (DRD4) gene polymorphism is associated with attachment disorganization in infants. *Molecular Psychiatry*, **5**, 633-637.

Lesch, K.P., Bengel, D., Heils, A., Sabol, S.Z., Greenberg, B.D., Petri, S., Benjamin, J., Muller, C.R., Hamer, D.H., & Murphy, D.L. (1996). Association of anxiety-related traits with a polymorphism in the serotonin transporter gene regulatory region. *Science*, **274**, 1527-1531.

Li, T., Xu, K., Deng, H., Cai, G., Liu, J., Liu, X., Wang, R., Xiang, X., Zhao, J., Murray, R.M., Sham, P.C., & Collier, D.A. (1997). Association analysis of the dopamine D4 gene exon III VNTR and heroin abuse in Chinese subjects. *Molecular Psychiatry*, **2**, 413-416.

Lichtenberg, P., Bachner-Melman, R., Gritsenko, I., & Ebstein, R.P. (2000). Exploratory association study between catechol-O-methyltransferase (COMT) high/low enzyme activity polymorphism and hypnotizability. *American Journal of Medical Genetics*, **96**, 771-774.

Lichter, J.B., Barr, C.L., Kennedy, J.L., Van Tol, H.H., Kidd, K.K., & Livak, K.J. (1993). A hypervariable segment in the human dopamine receptor D4 (DRD4) gene. *Human Molecular Genetics*, **2**, 767-773.

Lotta, T., Vidgren, J., Tilgmann, C., Ulmanen, I., Melen, K., Julkunen, I., & Taskinen, J. (1995). Kinetics of human soluble and membrane-bound catechol O-methyltransferase: a revised mechanism and description of the thermolabile variant of the enzyme. *Biochemistry,* **34**, 4202-4210.

Lyons, M.J., True, W.R., Eisen, S.A., Goldberg, J., Meyer, J.M., Faraone, S.V., Eaves, L.J., & Tsuang, M.T. (1995). Differential heritability of adult and juvenile antisocial traits. *Archives of General Psychiatry,* **52**, 906-915.

Malhotra, A.K., & Goldman, D. (2000). The dopamine D(4) receptor gene and novelty seeking. *American Journal of Psychiatry,* **157**, 1885-1886.

Malhotra, A.K., Virkkunen, M., Rooney, W., Eggert, M., Linnoila, M., & Goldman, D. (1996). The association between the dopamine D4 receptor (D4DR) 16 amino acid repeat polymorphism and novelty seeking. *Molecular Psychiatry,* **1**, 388-391.

Mannisto, P.T. (1998). Catechol O-methyltransferase: characterization of the protein, its gene, and the preclinical pharmacology of COMT inhibitors. *Advances in Pharmacology,* **42**, 324-328.

Manor, I., Eisenberg, J., Tyano, S., Sever, Y., Cohen, H., Ebstein, R.P., & Kotler, M. (2001). Family based association study of the serotonin transporter promoter-region polymorphism (5-HTTLPR) in attention deficit hyperactivity disorder (ADHD). *American Journal of Medical Genetics,* **105**, 91-95.

Martin, N.G., & Boomsma, D.I. (1989). Willingness to drive when drunk and personality: a twin study. *Behavior Genetics,* **19**, 97-111.

Mawson, A.R., Biundo, J.J., Jr., Clemmer, D.I., Jacobs, K.W., Ktsanes, V.K., & Rice, J.C. (1996). Sensation-seeking, criminality, and spinal cord injury: a case-control study. *American Journal of Epidemiology,* **144**, 463-472.

Mazzanti, C.M., Lappalainen, J., Long, J.C., Bengel, D., Naukkarinen, H., Eggert, M., Virkkunen, M., Linnoila, M., & Goldman, D. (1998). Role of the serotonin transporter promoter polymorphism in anxiety-related traits. *Archives of General Psychiatry,* **55**, 936-940.

McCourt, W.F., Gurrera, R.J., & Cutter, H.S. (1993). Sensation seeking and novelty seeking. Are they the same? *Journal of Nervous and Mental Disease,* **181**, 309-312.

McCown, W. (1989). The relationship between impulsivity, empathy and involvement in twelve step self-help substance abuse treatment groups. *British Journal of Addictions,* **84**, 391-393.

McCrae, R.R., & Costa, P.T., Jr. (1997). Personality trait structure as a human universal. *American Psychologist,* **52**, 509-516.

Mel, H., Horowitz, R., Ohel, N., Kramer, I., Kotler, M., Cohen, H., Gritsenko, I., & Ebstein, R.P. (1998). Additional evidence for an association between the dopamine D4 receptor (D4DR) exon III seven-repeat allele and substance abuse in opioid dependent subjects: Relationship of treatment retention to genotype and personality. *Addiction Biology,* **3**, 473-481.

Michel, G., Carton, S., & Jouvent, R. (1997). [Sensation seeking and anhedonia in risk taking. Study of a population of bungy jumpers]. *Encephale,* **23**, 403-411.

Mitsuyasu, H., Hirata, N., Sakai, Y., Shibata, H., Takeda, Y., Ninomiya, H., Kawasaki, H., Tashiro, N., & Fukumaki, Y. (2001). Association analysis of polymorphisms in the upstream region of the human dopamine D4 receptor gene (DRD4) with schizophrenia and personality traits. *Journal of Human Genetics,* **46**, 26-31.

Morgan, A.B., & Lilienfeld, S.O. (2000). A meta-analytic review of the relation between antisocial behavior and neuropsychological measures of executive function. *Clinical Psychology Review*, **20**, 113-136.

Murakami, F., Shimomura, T., Kotani, K., Ikawa, S., Nanba, E., & Adachi, K. (1999). Anxiety traits associated with a polymorphism in the serotonin transporter gene regulatory region in the Japanese. *Journal of Human Genetics*, **44**, 15-17.

Muramatsu, T., Higuchi, S., Murayama, M., Matsushita, S., & Hayashida, M. (1996). Association between alcoholism and the dopamine D4 receptor gene. *Journal of Medical Genetics*, **33**, 113-115.

Nigg, J.T., & Goldsmith, H.H. (1994). Genetics of personality disorders: perspectives from personality and psychopathology research. *Psychological Bulletin*, **115**, 346-380.

Niimi, Y., Inoue-Murayama, M., Murayama, Y., Ito, S., & Iwasaki, T. (1999). Allelic variation of the D4 dopamine receptor polymorphic region in two dog breeds, Golden retriever and Shiba. *Journal of Veterinary Medical Science*, **61**, 1281-1286.

Noble, E.P., Ozkaragoz, T.Z., Ritchie, T.L., Zhang, X., Belin, T.R., & Sparkes, R.S. (1998). D2 and D4 dopamine receptor polymorphisms and personality. *American Journal of Medical Genetics*, **81**, 257-267.

Okuyama, Y., Ishiguro, H., Nankai, M., Shibuya, H., Watanabe, A., & Arinami, T. (2000). Identification of a polymorphism in the promoter region of DRD4 associated with the human novelty seeking personality trait. *Molecular Psychiatry*, **5**, 64-69.

Ono, Y., Manki, H., Yoshimura, K., Muramatsu, T., Mizushima, H., Higuchi, S., Yagi, G., Kanba, S., & Asai, M. (1997). Association between dopamine D4 receptor (D4DR) exon III polymorphism and novelty seeking in Japanese subjects. *American Journal of Medical Genetics*, **74**, 501-503.

Osher, Y., Hamer, D., & Benjamin, J. (2000). Association and linkage of anxiety-related traits with a functional polymorphism of the serotonin transporter gene regulatory region in Israeli sibling pairs. *Molecular Psychiatry*, **5**, 216-219.

Paterson, A.D., Sunohara, G.A., & Kennedy, J.L. (1999). Dopamine D4 receptor gene: novelty or nonsense? *Neuropsychopharmacology*, **21**, 3-16.

Persson, M.L., Wasserman, D., Geijer, T., Frisch, A., Rockah, R., Michaelovsky, E., Apter, A., Weizman, A., Jonsson, E.G., & Bergman, H. (2000). Dopamine D4 receptor gene polymorphism and personality traits in healthy volunteers. *European Archives of Psychiatry and Clinical Neuroscience*, **250**, 203-206.

Pogue-Geile, M., Ferrell, R., Deka, R., Debski, T., & Manuck, S. (1998). Human novelty-seeking personality traits and dopamine D4 receptor polymorphisms: a twin and genetic association study. *American Journal of Medial Genetics*, **81**, 44-48.

Pomerleau, C.S., Pomerleau, O.F., Flessland, K.A., & Basson, S.M. (1992). Relationship of Tridimensional Personality Questionnaire scores and smoking variables in female and male smokers. *Journal of Substance Abuse*, **4**, 143-154.

Raine, A. (1996). Autonomic nervous system factors underlying disinhibited, antisocial, and violent behavior. Biosocial perspectives and treatment implications. *Annals of the New York Academy of Sciences*, **794**, 46-59.

Raine, A., Lencz, T., Bihrle, S., LaCasse, L., & Colletti, P. (2000). Reduced prefrontal gray matter volume and reduced autonomic activity in antisocial personality disorder. *Archives of General Psychiatry*, **57**, 119-127.

Raine, A., Venables, P.H., & Williams, M. (1990). Relationships between central and autonomic measures of arousal at age 15 years and criminality at age 24 years. *Archives of General Psychiatry,* **47**, 1003-1007.

Ricketts, M.H., Hamer, R.M., Sage, J.I., Manowitz, P., Feng, F., & Menza, M.A. (1998). Association of a serotonin transporter gene promoter polymorphism with harm avoidance behaviour in an elderly population. *Psychiatric Genetics,* **8**, 41-44.

Robbins, T.W., & Everitt, B.J. (1996). Neurobehavioural mechanisms of reward and motivation. *Current Opinion in Neurobiology,* **6**, 228-236.

Ronai, Z., Szekely, A., Nemoda, Z., Lakatos, K., Gervai, J., Staub, M., & Sasvari-Szekely, M. (2001). Association between Novelty Seeking and the -521 C/T polymorphism in the promoter region of the DRD4 gene. *Molecular Psychiatry,* **6**, 35-38.

Rubinstein, M., Phillips, T.J., Bunzow, J.R., Falzone, T.L., Dziewczapolski, G., Zhang, G., Fang, Y., Larson, J.L., McDougall, J.A., Chester, J.A., Saez, C., Pugsley, T.A., Gershanik, O., Low, M.J., & Grandy, D.K. (1997). Mice lacking dopamine D4 receptors are supersensitive to ethanol, cocaine, and methamphetamine. *Cell,* **90**, 991-1001.

Sander, T., Harms, H., Dufeu, P., Kuhn, S., Rommelspacher, H., & Schmidt, L.G. (1997). Dopamine D4 receptor exon III alleles and variation of novelty seeking in alcoholics. *American Journal of Medical Genetics,* **74**, 483-487.

Schroth, M.L. (1996). Scores on sensation seeking as a predictor of sexual activities among homosexuals. *Perceptual and Motor Skills,* **82**, 657-658.

Schultz, W. (1997). Dopamine neurons and their role in reward mechanisms. *Current Opinion in Neurobiology,* **7**, 191-197.

Seeger, G., Schloss, P., & Schmidt, M.H. (2001). Functional polymorphism within the promotor of the serotonin transporter gene is associated with severe hyperkinetic disorders. *Molecular Psychiatry,* **6**, 235-238.

Shields, P.G., Lerman, C., Audrain, J., Bowman, E.D., Main, D., Boyd, N.R., & Caporaso, N.E. (1998). Dopamine D4 receptors and the risk of cigarette smoking in African-Americans and Caucasians. *Cancer Epidemiological Biomarkers Prevention,* **7**, 453-458.

Strobel, A., Wehr, A., Michel, A., & Brocke, B. (1999). Association between the dopamine D4 receptor (DRD4) exon III polymorphism and measures of Novelty Seeking in a German population. *Molecular Psychiatry,* **4**, 378-384.

Strous, R.D., Bark, N., Parsia, S.S., Volavka, J., & Lachman, H.M. (1997). Analysis of a functional catechol-O-methyltransferase gene polymorphism in schizophrenia: evidence for association with aggressive and antisocial behavior. *Psychiatry Research,* **69**, 71-77.

Sullivan, P.F., Fifield, W.J., Kennedy, M.A., Mulder, R.T., Sellman, J.D., & Joyce, P.R. (1998). No association between novelty seeking and the type 4 dopamine receptor gene (DRD4) in two New Zealand samples. *American Journal of Psychiatry,* **155**, 98-101.

Tomitaka, M., Tomitaka, S., Otuka, Y., Kim, K., Matuki, H., Sakamoto, K., & Tanaka, A. (1999). Association between novelty seeking and dopamine receptor D4 (DRD4) exon III polymorphism in Japanese subjects. *American Journal of Medical Genetics,* **88**, 469-471.

Van Tol, H.H. (1996). The dopamine D4 receptor. *NIDA Research Monograph,* **161**, 20-38.

Van Tol, H.H. (1998). Structural and functional characteristics of the dopamine D4 receptor. *Advances in Pharmacology,* **42**, 486-490.

Vandenbergh, D.J., Rodriguez, L.A., Hivert, E., Schiller, J.H., Villareal, G., Pugh, E.W., Lachman, H., & Uhl, G.R. (2000). Long forms of the dopamine receptor (DRD4) gene VNTR are more prevalent in substance abusers: No interaction with functional alleles of the catechol-o-methyltransferase (COMT) gene. *American Journal of Medical Genetics,* **96**, 678-683.

Vandenbergh, D.J., Rodriguez, L.A., Miller, I.T., Uhl, G.R., & Lachman, H.M. (1997). High-activity catechol-O-methyltransferase allele is more prevalent in polysubstance abusers. *American Journal of Medical Genetics,* **74**, 439-442.

Vandenbergh, D.J., Zonderman, A.B., Wang, J., Uhl, G.R., & Costa, P.T., Jr. (1997). No association between novelty seeking and dopamine D4 receptor (D4DR) exon III seven repeat alleles in Baltimore Longitudinal Study of Aging participants. *Molecular Psychiatry,* **2**, 417-419.

von Knorring, L., von Knorring, A.L., Smigan, L., Lindberg, U., & Edholm, M. (1987). Personality traits in subtypes of alcoholics. *Journal of Studies of Alcohol,* **48**, 523-527.

Vukov, M., Baba-Milkic, N., Lecic, D., Mijalkovic, S., & Marinkovic, J. (1995). Personality dimensions of opiate addicts. *Acta Psychiatrica Scandinavica,* **91**, 103-107.

Weinshilboum, R., & Dunnette, J. (1981). Thermal stability and the biochemical genetics of erythrocyte catechol-O-methyl-transferase and plasma dopamine-beta-hydroxylase. *Clinical Genetics,* **19**, 426-437.

Wills, T.A., Windle, M., & Cleary, S.D. (1998). Temperament and novelty seeking in adolescent substance use: convergence of dimensions of temperament with constructs from Cloninger's theory. *Journal of Personality and Social Psychology,* **74**, 387-406.

Wise, R.A. (1996). Neurobiology of addiction. *Current Opinion in Neurobiology,* **6**, 243-251.

Young, T.J. (1990). Sensation seeking and self-reported criminality among student-athletes. *Perceptual and Motor Skills,* **70**, 959-962.

Zuckerman, M. (1994). *Behavioral expressions and biosocial bases of sensation seeking.* New York: Cambridge University Press.

PART C

GENETIC, ENVIRONMENTAL AND COGNITIVE FACTORS PREDISPOSING TO CRIME

Chapter 4

BEHAVIORAL GENETICS AND CRIME

Sharon S. Ishikawa, Ph.D. and Adrian Raine, D.Phil.
Department of Psychology, University of Southern California

Acknowledgments. This research was conducted with the support of a National Research Service Award (1 F32 MH12951-01) to the first author and an Independent Scientist Award (K02 MH01114) to the second author.

INTRODUCTION

One of the most hotly debated questions in the scientific arena has centered on whether criminal behavior has a genetic basis. Although it has been previously argued that genetics play no part in shaping antisocial and criminal behavior (e.g., Gottfredson & Hirschi, 1990), a growing literature base has served to substantiate that genetic factors are as important to the development of some forms of criminal activity as are environmental factors.

This attitudinal shift has occurred for several reasons. First, there are simply too many studies, in too many countries, using different methodologies that converge on the same conclusion: genes do play a role. Second, other, potentially less controversial fields of behavioral trait research have not only identified heritability in psychiatric disorders such as autism, schizophrenia and reading disability (Plomin, Owen, & McGuffin, 1994; Rutter, Silberg, O'Connor, & Simonoff, 1999b), but also in personality traits such as political conservatism (Crelia & Tesser, 1996). Thus, it would be surprising if criminal behavior—particularly recidivistic crime—was not in some way influenced by genetic factors.

Genetics of Crime: Clarifying Misconceptions

Before embarking on a summary of the research, it is important to clarify what behavior geneticists do *not* claim from their research on antisocial and criminal behavior. First, it is not claimed that there is a gene or specific set of genes that directly result in crime. Just as it is not assumed that there are specific genes that code for membership in the Republican or Democratic political parties, it is not assumed that there is a specific set of genes that directly code for criminal behavior. Rather, multiple genes—acting in combination—result in varying degrees of genetic *predisposition* to criminal behavior. This predisposition arises through heritable physiological processes such as neurotransmitter and autonomic nervous system functioning, which, in turn, predispose some individuals towards crime (Bohman, Cloninger, Sigvardsson, & von Knorring, 1982; Venables & Raine, 1987). A second, related point is that behavioral-genetic research does not attempt to explain why a specific person commits crime; heritability is a population concept and therefore cannot be applied at the individual level.

Third, criminal behavior is influenced both by environmental and genetic forces, as well as their interaction. Finding heritable influences does not imply that crime is entirely genetically determined. Adoption studies have repeatedly demonstrated the importance of factors such as adverse home environment, institutionalization early in life, multiple temporary placements, and low social class on criminal behavior (Cadoret & Cain, 1980; Cloninger & Gottesman, 1987; Crowe, 1974; Mednick & Finello, 1983; Rutter *et al.*, 1990). In addition, modern statistical approaches to the biometric analysis of twin data allow environmental variance to be partitioned into shared variance (i.e., common to both twins, such as parental divorce or death of parent) and non-shared variance (i.e., different for the two twins, such as different peer groups or differential treatment by parents). Studies of twins reared apart can also reveal important information on environmental protective factors that influence why one twin becomes criminal and the other genetically identical twin does not. Relatedly, in addition to the study of *additive* effects (i.e., environmental and genetic influences each explain some unique variance in criminal behavior), there may be an *interaction* between genes and environment. That is, some combination of genetic vulnerabilities and environmental risk factors may be more important to the expression of criminal behavior than are the separate, additive effects of genetic and environmental influences.

Finally, behavioral geneticists do not interpret their data to suggest that "crime is destiny." Environmental manipulations can be successful in reducing the incidence of crime by preventing full expression of genetic predispositional factors. For example, even very specific genetic disorders such as phenylketonuria (PKU), which results in mental impairment, can be prevented by environmental (i.e., dietary) interventions. Similarly, genetic influences for crime are seen as a *predisposition* to crime rather than as a *predestination* for crime.

RESEARCH ON THE GENETIC BASIS OF CRIME

Adult Twin Studies

Methodology

The twin method for ascertaining whether a given trait is to any extent heritable makes use of the fact that monozygotic (MZ) or "identical" twins are genetically identical, having 100% of their genes in common with one another. Conversely, dizygotic (DZ) or "fraternal" twins are less genetically alike than MZ twins, and are, in fact, no more genetically alike than non-twin siblings. Although many criminology texts (e.g., Ellis & Walsh, 2000; Sheley, 1991) state that DZ twins have, on average, 50% of their genes in common, this is not strictly true. All human beings have "fixed" or structural genes, as well as "polymorphic" genes. Fixed genes code for basic structure and make us all fundamentally similar (e.g., we all have two eyes, two arms, and a head), whereas polymorphic genes have different forms and determine things such as whether a person has blue eyes or brown eyes. The majority of genes are fixed, with the reader, irrespective of sex, having about 99% of his/her genes in common with the first author. In addition, fixed genes are shared across species to differing degrees, with both author and reader each sharing about 98% of their genes with chimpanzees (Britten, 1986). What determines individual variability *within* humans is the extent to which people differ on the final 1% of polymorphic genes. When DZ twins (or siblings) are said to have 50% of their genes in common, what is really meant is that they share 50% of this remaining 1% of genes. Furthermore, it should be remembered that all DZ twin pairs will not share exactly 50% of their genes in common. Some will be more genetically similar, some less so, with 50% representing an average that can vary.

When the trait being measured is a dichotomy (e.g., criminal/non-criminal), "concordance" rates are calculated for MZ and DZ twins separately. The current chapter will provide pairwise concordance rates, which refer to the likelihood that a pair of twins match on the behavioral trait of interest. For example, 70% concordance for crime in a set of MZ twins would mean that if one of the MZ pair is criminal, then the chance of the co-twin being criminal is 70%. Similar pairwise concordance rates can be calculated for DZ twins. If MZ twins have higher concordance rates for crime than DZ twins, then this provides some evidence that crime has a heritable component. Tetrachoric correlations can be calculated from each of these concordance rates, and the difference between these correlation coefficients, when doubled, gives an estimate of heritability, or the proportion of variance in criminality that can be attributed to genetic influences (Falconer, 1965). More complex, multivariate biometrical models have also been used more frequently in recent studies (e.g., Rutter, Silberg, O'Connor, & Simonoff, 1999a).

Empirical Findings

To date there appear to be 11 studies producing 14 analyses on the genetics of adult crime using twins (see Table 1). Studies vary widely in terms of the country of origin, age and sex composition, sample size, determination of zygosity, and definition of crime. With the exception of Coid, Lewis, and Reveley (1993), all analyses show greater pairwise concordance for criminality in MZ as opposed to DZ twins. If one averages concordance rates across all studies (weighting for sample sizes), these 14 analyses result in concordances of 44.0% for MZ twins and 21.6% for DZ twins, suggesting heritability for criminal behavior. With regard to the null findings of Coid *et al.*, this study was the only one using a sample selected for psychiatric illness (i.e., psychotic symptoms), which may limit its generalizability to criminal behavior.

Table 1. Pairwise concordance rates from twin studies of criminal behavior. N = number of pairs of twins. Summary figures are weighted for sample sizes.

Author	Location	Sex	MZ Twins % Concordance (*N*)	DZ Twins % Concordance (*N*)	Zygosity Assessment
Lange (1929)[a]	Bavaria	M	77 (13)	12 (17)	resemblance
Legras (1932)[a]	Holland	M	100 (4)	0 (5)	blood
Rosanoff *et al.* (1934)[a]	U.S.A.	M	76 (38)	22 (23)	resemblance
		F	26 (7)	25 (4)	
Stumpfl (1936)[a]	Germany	M	60 (15)	37 (19)	resemblance
		F	67 (3)	0 (2)	
Kranz (1936)[a]	Prussia	M	66 (32)	54 (43)	blood
Borgstrom (1939)[a]	Finland	M	75 (4)	40 (5)	blood
Slater (1953)	England	M	50 (2)	30 (10)	?
Yoshimasu (1961)[a]	Japan	M	61 (28)	11 (18)	blood
Dalgaard & Kringlen (1976)	Norway	M	26 (31)	15 (54)	blood
Cloninger & Gottesman (1987)[b]	Denmark	M	34 (73)	18 (146)	questionnaire
		F	20 (15)	7 (28)	
Coid *et al.* (1993)	United Kingdom	M/F	24 (92)	21 (109)	?
		TOTAL:	43.7 (357)	21.6 (483)	

[a] *Data obtained from Christiansen (1977a)*
[b] *The complete data were also analyzed by Carey (1992), and incomplete data were analyzed by Christiansen (1977b)*

One should note that researchers have identified several methodological limitations across these studies, with some suggesting that findings of heritability should be disregarded as a result. Mednick and Finello (1983) point out that two of the earlier studies (Stumpfl, 1936, Kranz 1936, as reported by Christiansen, 1977a) took place in Germany and Prussia during the Nazi era and may be politically biased. Interestingly, however, the MZ-DZ differences in these studies are small relative to some of the other studies, and in a detailed methodological analysis Christiansen (1977a) concluded that Kranz's work was the best of the pre-war twin studies on crime. In order to more systematically assess whether the total concordance rate was significantly affected by specific factors (e.g., gender, zygosity determination, small sample sizes, recent versus old studies), we re-calculated the total rate as a function of the moderator variables shown in Table 2. MZ/DZ concordance ratios were also calculated, with higher ratios indicating greater concordance in MZ twins relative to DZ twins.

Excluding the Coid *et al.* (1993) study resulted in MZ and DZ concordance rates of 50.5 and 21.8, respectively, and deleting the Germanic studies of Stumpfl (1936) and Kranz (1936) produced concordance rates of 40.3% (MZ) and 17.7% (DZ). Thus, the argument that results of all twin studies should be rejected on the grounds of political and, to a somewhat lesser extent, sampling bias, is attenuated. Twin studies have also been criticized due to a reliance on small sample sizes. However, when these studies were deleted, evidence for heritability for crime was still strong (42.7% for MZ, 21.7% for DZ). Similarly, twin studies have been criticized for method of zygosity assessment. However, as indicated in Table 2, substantial evidence for heritability was found in studies where zygosity was determined from blood samples (53.8% for MZ, 28.2% for DZ). Although overall concordance rates were lower for females than males (as would be expected given the substantially lower crime rates in females), the rates in female MZ twins were still more than three times that of female DZ twins (27.3 and 8.7, respectively). When looking only at the three most recent studies (Cloninger & Gottesman, 1987; Coid *et al.*, 1993; Dalgaard & Kringlen, 1976), concordance rates in the MZ and DZ twin sets were lower than for the total set of studies. However, they were still in the direction of MZ twins having higher concordance than DZ twins (27.5% for MZ and 17.58% for DZ), suggesting that results from the older twin studies should not necessarily be disregarded.

Table 2. Averaged pairwise concordance rates weighted by sample size for selected twin studies.

Selection Criterion	**MZ Twins % Concordance (N)**	**DZ Twins % Concordance (N)**	**MZ/DZ concordance Ratio**
Total Sample	44.0 (360)	21.6 (483)	2.03
Excluding Coid *et al.* (1993)	50.5 (265)	21.8 (374)	2.33
Excluding Stumpfl (1936) and Kranz (1936)	40.3 (307)	17.7 (419)	2.27
Excluding small samples (single digit studies)	42.7 (337)	21.7 (467)	1.98
Zygosity estimated by blood	53.8 (99)	28.2 (125)	1.90
Males only	52.9 (240)	23.1 (340)	2.30
Females only	27.3 (25)	8.7 (34)	3.14
Post-1975 only	27.5 (211)	17.6 (337)	1.56

However one views these data, the results appear fairly consistent that some degree of genetic predisposition for crime exists. Overall concordance ratios range from 1.5 to 3.1, indicating that even in the worst-case scenario, MZ twins are at least one-and-a-half times more likely to be concordant for crime than are DZ twins. While some scholars have attempted to reject results from all twin research on the basis of criticisms that affect only a select sample of studies, these criticisms have not been supported by the data presented above.

Limitations of Twin Studies

Twin studies nevertheless suffer from other methodological issues that limit the conclusions that can be drawn from individual studies. A common criticism of twin studies is that MZ twins may share a more common environment than DZ twins. For example, parents may treat MZ twins in a more similar fashion than DZ twins, thus artificially raising concordance rates in MZ twins. If this were true, the greater concordance for crime in MZ twins would actually reflect more of an environmental effect than a genetic one. Although there is some evidence to support this view (Allen, 1976), there are other methodological issues that can *decrease* estimates of heritability. For example, research suggests that greater genetic relatedness between a pair of twins leads to increased similarity in behavior and, in turn, this increased similarity in behavior elicits increased similarity in parental treatment (Kendler, 1983). In addition, some twins make attempts to "de-identify" or be different from one another (Schacter & Stone, 1985), or develop opposite role relationships (e.g., dominant-submissive; Moilanen, 1987).

Identical twinning may also result in non-genetic biological differences that can accentuate phenotypic differences. For example, there can be a greater discrepancy in the birth weights of MZ twins relative to DZ twins due to the foeto-foetal transfusion syndrome, and certain perinatal complications are more likely to occur in MZ than DZ twins (Rutter *et al.*, 1997). Because birth complications have been linked to differences in behavior and cognition, these non-genetic biological factors could result in an exaggeration of behavioral differences between MZ twins, thus reducing heritability estimates.

Overall, then, methodological problems in twin studies are just as likely to decrease heritability estimates as they are to inflate them. Rutter *et al.* (1990) have suggested that these effects likely cancel each other out. Thus, in spite of its limitations, the study of twins remains a key method in the field of behavior genetics, and future twin studies of crime using up-to-date biometric modeling are likely to make increasingly important contributions to genetic research on crime.

Identical Twins Reared Apart

Studies of identical twins reared apart circumvent the main criticism of twin studies that MZ twins share a common environment in addition to a common genetic make-up. Although this is a powerful method for establishing heritability for a trait, very few studies have been conducted due to the low frequency of such adoption placements. Grove *et al.* (1990) studied 32 sets of monozygotic twins who were separated and reared apart shortly after birth, with zygosity assessed using a number of techniques (including blood assessment). A continuous score for self-reported antisocial behavior in both childhood and adulthood was derived from a standardized interview schedule. Statistically significant heritabilities were obtained for antisocial behavior in both childhood (0.41) and adulthood (0.28).

These findings are consistent with anecdotal evidence from twins reared apart. In a report of eight case studies of MZ twin pairs in which it was known that one twin was registered for criminality, four were concordant for crime (Christiansen, 1977a). Similarly, one of the four concordant cases reported by Schwesinger (1952) describes a pair of MZ twins separated at the age of 9 months and brought up by different sets of parents described as having dissimilar personalities. At age 9 one twin was moved and raised in a town, while the other twin remained in the desert. Independently, both twins started to commit juvenile crimes after puberty and were separately incarcerated. Thus, studies of twins reared apart clearly represent an important research strategy for understanding the genetics of crime, but to date the data are limited. While the case studies are not methodologically strong, they do illustrate the usefulness of this approach and, together with the methodologically stronger study of Grove *et al.* (1990), add support for the notion that there is a genetic predisposition to crime and antisocial behavior in some individuals.

Adoption Study Methodology

In adoption studies, offspring are separated from their criminal, biological parents early in life and adopted by other families. If these offspring grow up to commit crime at greater rates than adopted children whose biological parents were not criminal, this would indicate a genetic influence with its origin in the subject's biological parents. This research paradigm provides an additional strategy for separating out genetic and environmental influences, although the issue of selective adoption placement (i.e., adoptive parents are typically screened for antisocial behavior) may limit the ability to detect environmental effects (Rydenour, 2000).

A variation of this type of study is the "cross-fostering" paradigm, which has been used most extensively in experimental genetic studies with animals. When applied to studies with humans, the offspring whose biological parents are criminal or non-criminal are adopted by parents who themselves are either criminal or non-criminal. This 2 x 2 design capitalizes on what is effectively a natural experiment and allows for a more systematic exploration of genetic and environmental influences. As will be seen later, this paradigm also allows for an assessment of possible interactions between genetic and environmental influences, and sidesteps the issue of selective adoption placement.

Evidence from Adoption Studies

In total, 15 well-executed adoption studies on antisocial behavior have been conducted in Denmark, Sweden, and the United States. The main results of these studies are tabulated in Table 3 and grouped under 5 different headings corresponding to the different adoption study from which they were derived. It should be noted that within each of the five headings, some studies are based on analyses of essentially the same or enlarged data set. For example the interaction between genetics and environment in males found by Cloninger, Sigvardsson, Bohman and von Knorring (1982) and the heritability for petty crime in males found by Bohman, Cloninger, Sigvardsson, and von Knorring (1982) are based on the same data set. Importantly, however, these studies address different issues on the genetics of crime. Moreover, studies under one heading are independent of studies under another heading.

A review of Table 3 reveals that all studies except one (Bohman, 1978) support the hypothesis that there is some genetic predisposition to crime. This basic finding appears robust based on the fact that positive results were found by several *independent* research groups. Similarly, heritability for crime has been found in several different countries (United States, Denmark, Sweden), suggesting initial cross-cultural generalizability. However, future studies in non-Western countries are required to further establish the generalizability of the findings.

Table 3. Key findings from adoption studies of adult criminal behavior.

Danish Adoption Study	**Main Finding**
Mednick *et al.* (1984)	Significant correlation between adoptees and biological parents for property but not violent crimes
Baker (1986)	Heritable influences for property crimes and antisocial disorders
Moffitt (1987)	Mental illness and crime in biological parents increases risk of recidivistic, non-violent crime in adopted away offspring
Silverton (1988)	Criminal behavior genetically linked to schizophrenia
Baker *et al.* (1989)	Convicted females more genetically predisposed to crime than convicted males
Swedish Adoption Study	**Main findings**
Bohman (1978)	No link between general crime in biological fathers and crime adoptees
Bohman *et al.* (1982)	Heritability for petty crime in males
Cloninger *et al.* (1982)	Interaction between genetic and environmental predispositions for petty crime in males
Sigvardsson *et al.* (1982)	Heritability for petty crime in males replicated in females; female petty crime requires stronger genetic predisposition
Iowa 1974 Adoption Study	**Main Finding**
Crowe (1974)	Higher rates of antisocial personality in adopted away offspring of female offenders
Iowa 1978 Adoption Study	**Main Finding**
Cadoret (1978)	Heritability for adult antisocial behavior
Cadoret & Cain (1980)	Antisocial personality or alcoholism in biological parent predicts antisocial behavior in adolescent adoptee
Cadoret *et al.* (1985)	Heritability for antisocial personality
Cadoret *et al.* (1987)	Heritability for adult antisocial behavior and personality
Missouri Adoption Study	**Main Finding**
Cadoret *et al.* (1983)	Interaction between genetic and environmental predispositions for adolescent antisocial behavior

In addition to the traditional adoptee design, at least one cross-fostering adoption study has been conducted (Mednick, Gabrielli, &

Hutchings, 1984). These researchers based their analyses on 14,427 adoptions that took place in Denmark between 1927 and 1947. Infants were adopted out immediately in 25.3% of cases, 50.6% within one year, 12.8% in the second year, and 11.3% after age 2. Court records were obtained on 65,516 biological parents, adoptive parents, and adoptees in order to assess which subjects had convictions. All of the findings presented below were independent of the socioeconomic status of the adoption homes and of the age at which the infant was taken away from the biological mother.

When both adoptive and biological parents were non-criminal (i.e., neither genetic nor environmental predispositions present), 13.5% of the adoptees had a criminal record. This increased to 14.7% when only the adoptive parents were criminal (i.e., environmental predisposition only). However, when only the biological parents were criminal (i.e., genetic predisposition only), the conviction rate in the adoptees increased to 20.0%, and when both adoptive and biological parents were criminal (i.e., genetic and environmental predispositions both present), the conviction rate increased to 24.5%. The effect of having a biological parent who was criminal, compared to having neither biological nor adoptive parents who were criminal, resulted in a statistically significant increase in the likelihood of the adoptee becoming criminal.

In addition, Mednick *et al.* (1984) found a statistically significant relationship between number of convictions in the biological parents and number of convictions in the adoptees. The more criminal convictions the biological parent had, the more convictions the adoptee was likely to have, suggesting a genetic influence on crime. However, this effect was only observed for property offenses; no significant effect was observed for violent offenses. Similar findings for the heritability of petty, property crimes but not of violent crimes were observed by Bohman *et al.* (1982) and Sigvardsson Cloninger, Bohman and von Knorring (1982).

This lack of findings for violent crime is not to say that aggression has no heritable basis. In an updated analysis of extensive twin data collected by Christiansen (1977b), Cloninger and Gottesman (1987) estimated heritability to be .78 for property crimes versus .50 for crimes against persons. This implies that while twin data suggest heritability for violence, the heritability for non-violent offending is much higher than it is for violent offending. Interestingly, Cloninger and Gottesman also found that the genetic predisposition for property offending was unrelated to the genetic predisposition for violent offending. Thus, separate genetic influences operate for these two types of criminal behavior.

There may be reasons why heritability exists for violence but is not able to be observed. The lower base rates for violence reduces statistical power and, thus, the ability to detect a significant effect. In addition, specific forms of violence (e.g., repeated assault) may have heritable factors while others do not (Carey, 1989). Carey (1989) has also suggested that studies using official records may have inaccuracies lowering heritability, although it is unclear why inaccurate records would affect violent offending but not property offending.

The negative findings for criminal violence must also be set against the strikingly positive findings for general aggressive behavior, that is, measures of aggression that do not rely on official criminal conviction. Dozens of studies show that aggressive behavior in adults and children is influenced by genetic factors (e.g., Coccaro, Bergeman, Kavoussi, & Seroczynski, 1997; Rushton, Fulker, Neale, Nias, & Eysenck, 1986; Vernon, McCarthy, Johnson, Jang, & Harris, 1999). In addition, a meta analysis conducted on 24 individual studies of aggressive behavior in male and female adults and children found that 50% of the variance in aggression could be attributed to genetic factors (Miles & Carey, 1997). One particularly interesting finding was that the heritability for aggression changed over time. While genes and common environment were equally important in influencing aggression in childhood, in adulthood the role of heritability increased and the role of common environment became negligible. Heritability was also slightly higher for male aggression than female aggression. Importantly, this set of findings was not attributable to methodological artifacts (e.g., selective adoption placements), and highlight the importance of expanding the research from a focus on crime to a more general focus on antisocial behavior.

Juvenile Twin Studies of Crime

Twin studies of juvenile crime are listed in Table 4. Previous reviews of juvenile crime (Carey, 1989; Eysenck, 1977; Rutter *et al.*, 1990; Trasler, 1987) argue that there is little evidence for genetic influences. Broadly speaking these conclusions are accurate, although the possibility of genetic liability should not be completely dismissed. Summary figures weighted by sample size across the first seven studies yield pairwise rates of 84.3% for MZ and 68.6% for DZ juvenile twins. Although the MZ and DZ concordance rates for juvenile crime are each much higher than for adult crime, the smaller difference between the MZ and DZ concordance rates suggests only a modest genetic effect. Thus, while there is some evidence for heritability of criminal behavior among children, the data suggest that environmental effects play a more substantial role.

Table 4. Pairwise concordance rates from twin studies of juvenile crime. N = number of pairs of twins. Total figures are weighted by sample size.

Author	Location	Sex	% MZ concordance (N)	% DZ concordance (N)
Rosanoff (1934)[a]	U.S.A.	M	100 (29)	71 (17)
Rosanoff (1934)[a]	U.S.A.	F	92 (12)	100 (9)
Kranz (1937)[a]	Prussia	M/F	69 (16)	59 (22)
Sugmati (1954)[a]	Japan	?	80 (10)	20 (5)
Mitsuda (1961)[a]	Japan	?	73 (15)	75 (4)
Hayashi (1967)[a]	Japan	M	80 (15)	75 (4)
Shields (1977)	England	M	80 (5)	78 (9)
Lyons (1996)	U.S.A.	M	47[b] (231)	41[b] (218)
	TOTAL[c]		84.3 (102)	68.6 (70)

[a] *Data obtained from Christiansen (1977a)*

[b] *Probandwise concordance rate*

[c] *Value calculated using only pairwise concordance rates; Lyons (1996) is excluded.*

GENETIC BASIS OF CRIME-RELATED BEHAVIORAL CONSTRUCTS

Childhood Conduct Disorder and Antisocial Behavior

In more recent years, genetics research has shifted from examining juvenile crime to examining childhood and adolescent antisocial, aggressive behavior. Table 5 outlines findings from several studies where antisocial behavior is not defined according to official criminal conviction but to self- or parental reports. With the exception of Bohman (1972) and Cadoret (1978), which both relied on official measures of criminality in the biological parent, all studies found significant heritability for delinquent, aggressive behavior. This discrepancy with studies of juvenile crime emphasizes the importance of studying the underlying behavioral traits that likely contribute to the legally defined construct of crime.

Table 5. Additional twin studies of antisocial behavior in children and juveniles.

Author	Main Finding
Bohman (1972)	No excess of juvenile delinquency (truancy, lying, stealing) in 10 and 11 year old adoptees in adopted away children whose biological parents were criminal.
Cadoret (1978)	No evidence for heritability for child diagnosis of conduct disorder (measured by psychiatric interview) in a sample of 10-17 year-old child adoptees whose biological parents were criminal.
O'Connor *et al.* (1980)	Evidence for heritability for "bullying" as rated by mothers (Holzinger's H = .60) in 54 MZ and 33 DZ 7.6 year olds.
Cadoret *et al.* (1983)	Evidence for significant H x E interaction in three samples of adoptees for adolescent antisocial behavior (measured by mothers' report of antisocial behaviors). No evidence, however, in two of these samples for a main effect of heredity using a multiple regression model.
Rowe (1983)	Significant heritability for self-report delinquency in both male (H = .36) and female (H = .52) children in grades 8-12. Heritability as high as .70 suggested by biometrical genetic modeling.
Jarey & Stewart (1985)	Adopted away offspring with aggressive conduct disorder were more likely to have biological parents with antisocial personality disorder.
Ghodsian-Carpey & Baker (1987)	Heritabilities ranging from .24 to .94 for mothers' ratings of different types of aggressive behavior in 21 MZ and 17 DZ 4-7 year twins.
Lytton *et al.* (1988)	Heritability of .44 in 9-year old male twins using the Child Behavior Questionnaire.
Stevenson & Graham (1988)	Heritability of .42 in 13 year old male twins assessed on the Child Behavior Questionnaire. No heritability in females.
Grove *et al.* (1990)	Identical twins reared apart: child antisocial behavior before the age of 15 assessed from a clinical interview yielded a heritability of 0.41.
Cadoret *et al.* (1995)	Adoptees with biological parents diagnosed with

	antisocial personality disorder were more likely to have conduct disorder than those with non-antisocial biological parents.
Ge *et al.* (1996)	Adoptees more likely to antisocial/ hostile behavior if biological parent had antisocial personality/substance abuse.
Slutske *et al.* (1997)	Found 71% heritability for conduct disorder among 2,682 pairs of Australian twins. Heritability estimate was largely the same for males and females.
Simonoff *et al.* (1998)	Child and maternal ratings in twin study of 8-26 year-olds both suggest heritability of property offenses and aggression. However, child ratings suggested general liability to antisocial/conduct disordered behavior, whereas maternal ratings were more specific to conduct behavioral problems and not as strongly related to status violations and oppositional behavior. Maternal ratings also suggested greater effect of common environment.
Eley *et al.* (1999)	Parent ratings indicated greater genetic influence on aggressive than non-aggressive antisocial behavior in 8-16 twin pairs. Shared environment significantly affected non-aggressive behavior. Shared correlation between aggressive and non-aggressive behavior was accounted for by both genetic and environmental factors for boys, but primarily by genetic factors for girls.
Hudziak *et al.* (2000)	Parental ratings provided estimates of genetic effects on aggression ranging from 70-77% in 8-12 year old twins.

In addition, although Table 5 summarizes the heritability findings, it should be pointed out that most of the studies found significant main effects for environmental factors. Moreover, Cadoret, Cain and Crowe (1983) found that antisocial behavior was accounted for by an interaction between genetic and environmental factors; main effects were not observed. Thus, as with criminal behavior in adults, antisocial behavior in children and adolescents is determined both by genetic and environmental factors. In contrast to the adult literature, however, the heritability estimates for antisocial behavior in children and adolescents are somewhat lower than that for adults (Miles & Carey, 1997; Rutter, 1996).

Findings of decreased heritability of early antisocial behavior may result from the fact that most children who are antisocial in childhood do not go on to become criminal, whereas many adult criminals are antisocial in childhood and adolescence. Drawing upon Moffitt's (1993) distinction between life-course persistent offending and adolescent-limited offending, it is possible that environmental influences—such as delinquent peer groups and other social factors—shape the behavior of those who limit their antisocial behavior to adolescence. In contrast, genetic factors may play a relatively greater role in shaping the behavior of those who begin aggressing early in childhood and persist into adulthood. Unfortunately, however, this prediction does not appear to have been empirically tested. A related explanation for the reduced heritability of early antisocial behavior is that peer-group pressures are particularly strong in adolescence, regardless of genetic predisposition. As such, environmental factors may play a much stronger role in juvenile delinquency than genetic factors.

An alternative reason as to why heritability of antisocial behavior is weaker and environmental influence is greater at younger ages may be that time is needed for the genetic mechanisms underlying antisocial behavior to be fully expressed. Data from a twin study on the inheritance of intelligence by Wilson (1983) provide a useful heuristic parallel. Wilson (1983) found that MZ twins became increasingly concordant for IQ from age 3 months to 15 years, whereas DZ twins regressed to intermediate levels of concordance by age 15 years. This greater heritability for IQ at later ages suggests that genetic mechanisms underlying intelligence are dependent on the full development of brain maturational processes that unfold in the context of the individual's living environment. In much the same way, genetic and biological mechanisms underlying antisocial behavior may require full maturation of the organism to achieve maximal expression. For example, even in an ideal rearing environment, the pre-frontal areas of the brain do not reach full developmental maturity until early adulthood, and poor functioning of this brain structure has been implicated in criminal and violent offenders (Kuruoglu *et al.*, 1996; Raine, Lencz, Bihrle, LaCasse, & Colletti, 2000; Raine *et al.*, 1998).

The Interplay of Genes and Environment

As mentioned throughout this chapter, evidence of genetic liability for antisocial behavior does not preclude the importance of environmental effects. With the exception of Ge *et al.* (1996), however, genetic and environmental effects in the studies reviewed above have largely been evaluated as independent influences on the development of antisocial or criminal behavior. A crucial question that remains is whether heredity and the environment interact with each other to evoke antisocial behavior. Studies that have explored this issue are reviewed below.

Gene and Environment Interactions

A gene x environment interaction suggests that, when certain environmental and genetic risk factors occur together, then an individual is most likely to manifest criminal behavior. Moreover, a significant interaction between heredity and environment means that the interaction term accounts for more variance in criminal behavior than either influence alone.

Evidence for an interaction between genetics and environment in crime comes from several studies. For example, Cloninger *et al.* (1982) conducted a cross-fostering analysis of "petty" criminality (i.e., infrequent, non-alcohol-related, minor property offenses). Swedish adoptees (n = 862) were divided into four groups depending on the presence or absence of (a) congenital predisposition (i.e., whether biological parents were criminal) and (b) post-natal predisposition (i.e., measures of child rearing experiences, adoptive placement). When both heredity and environmental predisposition factors were present, 40% of the adoptees were criminal. However, when only congenital factors or only post-natal factors were present, 12.1% and 6.7% of the adoptees, respectively, were criminal. In addition, when both factors were absent, only 2.9% of the adoptees were criminal. The 40% criminality rate when biological and environmental factors were both present is greater than simply adding the "congenital only" and "postnatal only" conditions together, thus indicating a non-additive interaction between the two factors. Similarly, Cadoret *et al.* (1983) presented data from three adoption studies that provided evidence for gene x environment interactions. In addition, Cadoret, Yates, Troughton, Woodworth and Stewart (1995) found that increased aggression in male and female offspring was significantly higher when the offspring had biological parents with antisocial personality and an adverse adoptive home environment. Crowe (1974) also found a trend for a gene x environment interaction in his analysis of adopted-away offspring of female prisoners.

Despite the compelling nature of these studies, some research limitations should be addressed. In the study by Cloninger *et al.* (1982), the cell size in the critical category (i.e., both heredity and environmental factors present) was small (n = 10) compared to the other categories (n = 66, 120, 66). However, Cloninger and Gottesman (1987) later found a similar gene x environment interaction in a larger sample of females from the same twin study data set, suggesting some stability of the finding.

A more important problem than small sample size is that some studies simply fail to find evidence for such interactions. For example, Cloninger, Christiansen, Reich and Gottesman (1978) concluded that their study of relatives of 60 probands with felony convictions "...raise serious doubt about the importance of nonlinear gene-environment interaction. Our analyses support additive or linear models without major interactions" (p. 949). It has been argued, however, that analysis of variance (ANOVA) is not particularly sensitive to detecting interactions, thus suggesting that the failure to find non-additive effects is a result of statistical considerations rather than conceptual issues (Wahlsten, 1990). Given the mixed results in the few studies testing an interactive effect, this issue warrants further scrutiny.

Gene-Environmental Correlations

Another important concept in the field of behavioral genetics is that of gene-environment correlations. A gene-environment correlation addresses the extent to which certain genotypes are selectively exposed to certain environments (Scarr & McCartney, 1983). For example, with regard to antisocial and criminal behavior, antisocial parents are more likely to provide a risky home environment than are non-antisocial parents. Similarly, children with behavior problems are more likely to elicit negative parenting behaviors than are children without such problems. Thus, treating genetic and environmental influences as independent may be artificial because part of the genetic influence could be mediated through the resulting environment (Rutter *et al.*, 1999b).

An interesting example of such a correlation comes from a study by Ge *et al.* (1996). The authors found that (a) antisocial behavior in the biological parents was associated with increased negative parenting in the adoptive parents, and that (b) the presence of antisocial personality and/or substance abuse in biological parents was associated with increased antisocial behavior in the adopted away child. Thus, the association between negative parenting in the adoptive parent and antisocial behavior in the child was mediated, in part, by genetic processes. Similar findings have been found by O'Connor, Deater-Deckard, Fulker, Rutter and Plomin (1998). In addition, Meyer *et al.* (2000) found that the influence of parental conduct disorder on offspring conduct disorder was mediated, in part, through marital discord. One of the goals of future behavior-genetic studies should be to further examine the interplay between genes and environment in this fashion.

Biological Markers for Antisocial and Criminal Behavior

One important contribution that genetic research on crime can make is in helping to identify the biological mechanisms that shape antisocial and criminal behavior (Raine, Venables & Williams, 1990; Rutter, 1996). Many studies have clearly demonstrated that there is a biological basis to antisocial and violent behavior, including factors such as hormone levels (Archer, 1991; Susman, Dorn, & Chrousos, 1991), poor functioning of the prefrontal region of the brain (Raine *et al.*, 1998), various other neurological deficits (Elliott, 1988), low physiological arousal (Raine *et al.*, 1990), and large body build (Wilson & Herrnstein, 1985). Within this broad field, perhaps the two best replicated biological markers of antisocial, aggressive behavior are low resting heart rate and low levels of serotonin. Findings from these areas, as well as preliminary findings from work with other neurotransmitters, are briefly reviewed below. More extensive coverage of low resting heart rate can be found in Chapter 8 of this volume; for a detailed analysis of serotonergic metabolism and aggression, see Chapter 9.

Cardiovascular Underarousal

Low heart rate is an indicator of autonomic underarousal that has substantial heritability. A meta-analysis of 29 studies yielding 38 effect sizes on resting heart rate and antisocial and aggressive behavior in childhood (Raine, 1997) produced an overall unweighted effect size (d) of 0.71 and a sample-corrected effect size of 0.56 ($p < .00001$). Such values are regarded as medium to large effect sizes (Cohen, 1988). The 15 independent studies of cross-sectional, community-based studies produced a sample-corrected d of 0.83 ($p < .0001$), which is regarded as a large effect size. In addition to cross-sectional findings, low resting heat rate measured during adolescence has been shown to predict crime in adulthood (e.g., Raine *et al.*, 1990; Wadsworth, 1976).

It has been hypothesized that low heart rate relates to aggressive antisociality because it reflects fearlessness (Raine, 1993). Individuals who lack fear are thought to be more likely to engage in aggressive encounters because they do not fear physical harm and/or because they do not condition to punishment (i.e., they do not respond to the negative aspects of fighting, such as injury and social punishment). A second explanation is that individuals with low arousal (i.e., low heart rate) seek out stimulation to increase their physiological arousal to more optimal levels, with one pathway involving aggressive, criminal, and other antisocial behaviors (Eysenck, 1997; in addition, see Chapter 2 for a more detailed discussion of sensation-seeking and psychopathy).

Serotonergic Dysfunction

Neurotransmitters within the brain are the "chemical messengers" responsible for regulating most of the brain's functions. With regard to the neurotransmitter serotonin (5-HT), an inverse relationship with aggressive/violent behavior has been fairly consistently observed across studies (Coccaro & Kavoussi, 1996; Linnoila & Virkkunen, 1992; Miczek, Weerts, Haney, & Tidey, 1994). Markers of low serotonin are associated with aggression in a variety of subpopulations such as naval recruits (Brown *et al.*, 1982; Brown, Goodwin, Ballenger, Goyer, & Major, 1979), incarcerated juvenile offenders (Unis *et al.*, 1997), incarcerated murderers diagnosed with Antisocial Personality Disorder (O'Keane *et al.*, 1992), aggressive psychiatric patients (Coccaro & Kavoussi, 1996; Stanley *et al.*, 2000), and depressed persons with a history of suicide attempts or completions (Coccaro & Kavoussi, 1996). Consistent with individual studies, a recent meta-analysis found that the overall CSF serotonin-antisocial behavior relationship is statistically significant and moderate in magnitude (Moore, Scarpa & Raine). In addition, genetics research has recently identified the serotonin transporter gene, in particular, as being important to the study of aggressive behavior (Suomi, 2000).

Scientists postulate that low 5-HT predisposes an individual to engage in aggressive behavior by adversely affecting impulse control (Coccaro & Kavoussi, 1996; Linnoila & Virkkunen, 1992). Support for this position comes from studies finding lower 5-HT in recidivistic rather than one-time violent offenders (Linnoila *et al.*, 1983), impulsive rather than premeditated violent offenders (Linnoila *et al.*,

1983; Virkkunen *et al.*, 1994), and behaviorally impulsive psychiatric patients rather than non-aggressive controls (Fava *et al.*, 2000; Hibbeln *et al.*, 1998). Abnormal serotonergic function has also been associated with other syndromes of behavioral disinhibition (i.e., attention deficit hyperactivity disorder) that have been associated with increased antisocial behavior (Comings, 2000). Interestingly, analog studies with monkeys have also found that, while average or above-average serotonin corresponds with on-going social competence, abnormally low serotonin both increases mortality through exposure to dangerous, aggressive behavior (Higley, Mehlman, *et al.*, 1996), and decreases social dominance ranking (Higley, King, *et al.*, 1996).

Other Neurotransmitter Correlates

In addition to serotonin, studies of monoamine oxidase and dopamine genes have shown an association with aggressive, antisocial behavior. In a fascinating study of four generations of a Dutch family, Brunner, Nelen, Breakefield, Ropers and van Oost (1993) found that each male subject characterized by impulsive aggression had a point mutation in the eighth exon on the monoamine oxidase-A structural gene. While Brunner's study is clearly only a beginning for the study of monoamine oxidase and aggression, it is consistent with the overall evidence for the heritability of aggressive, antisocial behavior. In addition, although the empirical literature does not support a general dopamine-antisocial behavior relationship (Raine, 1993), abnormal levels of dopamine have been found in select patient populations such as alcohol abusers (Kuikka *et al.*, 1998; Tiihonen *et al.*, 1995) and aggressive individuals with Alzheimer's Disease (Sweet *et al.*, 1998). Genetic studies on dopamine receptor genes (DRD1 – DRD5) and the dopamine transporter gene also reveal that certain gene variants are significantly related to impulsive, compulsive, addictive, and aggressive behavior problems (Comings, 2000). Dopamine is thought to be related to aggressive and antisocial behavior through its role in the initiation and execution of behavior and in behavioral reinforcement (Miczek, Haney, Tidey, Vivian, & Weerts, 1994). However, the inconsistencies in the literature warrant further investigation in order to elucidate the specific association between dopamine and antisocial behavior (Goldman & Fishbein, 2000).

IMPLICATIONS AND FUTURE DIRECTIONS OF GENETICS RESEARCH

Advances in genetics research and completion of the human genome project have provided compelling evidence that heredity influences complex behaviors such as antisocial and aggressive behavior (Plomin *et al.*, 1994). With this technological capacity, however, comes social and moral responsibility. Research on the genetic liability for antisocial behavior, for example, may have implications for individuals involved in criminal proceedings (Goldman & Fishbein, 2000), or may be used to inform social policy affecting large groups of people. However, one must remember that heritability is a population concept and cannot be applied to any single individual. Moreover, heritable, complex behaviors are significantly shaped by environmental factors; identification of a specific gene polymorphism is not meaningful when evaluated outside of the context of the individual's behavior and environment. Thus, while studies on the genetics of complex human behavior may ultimately help reduce crime through the development of effective treatment and prevention strategies (Comings, 2000; Goldman & Fishbein, 2000), such interventions must take into account the complex, reciprocal interaction between genes and the environment.

REFERENCES

Allen, G. (1976). Scope and methodology of twin studies. *Acta Geneticae Medicae Gemellologiae,* **25**, 79-85.

Archer, J. (1991). The influence of testosterone on human aggression. *British Journal of Psychology*, **82**, 1-28.

Baker, L.A. (1986). Estimating genetic correlations among discordant phenotypes: An analysis of criminal convictions and psychiatric hospital diagnoses in Danish adoptees. *Behavior Genetics*, **16**, 127-142.

Baker, L.A., Mack, W., Moffitt, T.E., & Mednick, S.A. (1989). Etiology of sex differences in criminal convictions in a Danish adoption cohort. *Behavior Genetics,* **19**, 355-370.

Bohman, M. (1972). A study of adopted children, their background, environment, and adjustment. *Acta Psychiatrica Scandinavica*, **61**, 90-97.

Bohman, M. (1978). Some genetic aspects of alcoholism and crime: A population of adoptees. *Archives of General Psychiatry*, **35**, 269-276.

Bohman, M., Cloninger, R., Sigvardsson, S., & von Knorring, A.L. (1982). Predisposition to petty criminality in Swedish adoptees. I. Genetic and environmental heterogeneity. *Archives of General Psychiatry*, **39**, 1233-1241.

Britten, R.J. (1986). Rates of DNA sequence evolution differ between taxonomic groups. *Science*, **231**, 1393-1398.

Brown, G.L., Ebert, M.H., Goyer, P.F., Jimerson, D.C., Klein, W.J., Bunney, W.E., & Goodwin, F.K. (1982). Aggression, suicide, and serotonin: Relationships to CSF amine metabolites. *American Journal of Psychiatry*, **139**, 741-746.

Brown, G.L., Goodwin, F.K., Ballenger, J.C., Goyer, P.F., & Major, L.F. (1979). Aggression in humans correlates with cerebrospinal fluid amine metabolites. *Psychiatry Research*, **1**, 131-139.

Brunner, H.G., Nelen, M., Breakefield, X.O., Ropers, H.H., & van Oost, B.A. (1993). Abnormal behavior associated with a point mutation in the structural gene for monoamine oxidase A. *Science*, **262**, 578-580.

Cadoret, R.J. (1978). Psychopathology in adopted-away offspring of biological parents with antisocial behavior. *Archives of General Psychiatry*, **35**, 176-184.

Cadoret, R., & Cain, C. (1980). Sex differences in predictors of antisocial behavior in adoptees. *Archives of General Psychiatry*, **37**, 1171-1175.

Cadoret, R.J., Cain, C.A., & Crowe, R.R. (1983). Evidence for gene-environment interaction in the development of adolescent antisocial behavior. *Behavior Genetics*, **13**, 301-310.

Cadoret, R.J., O'Gorman, T.W., Troughton, E., & Heywood, E. (1985). Alcoholism and antisocial personality: Interrelationships, genetic, and environmental factors. *Archives of General Psychiatry*, **42**, 161-167.

Cadoret, R.J., Troughton, E., & O'Gorman, T.W. (1987). Genetic and environmental factors in alcohol abuse and antisocial personality. *Journal of Studies on Alcohol*, **48**, 1-8.

Cadoret, R.J., Yates, W.R., Troughton, E., Woodworth, G., & Stewart, M.A. (1995). Genetic-environmental interaction in the genesis of aggressivity and conduct disorders. *Archives of General Psychiatry*, **52**, 916-924.

Carey, G. (1989). *Genetics and violence: Human studies*. Commissioned paper for Panel on the Understanding and Control of Violent Behavior. Washington: National Academy of Sciences.

Carey, G. (1992). Twin imitation for antisocial behavior: Implications for genetic and family environment research. *Journal of Abnormal Psychology*, **101**, 18-25.

Christiansen, K.O. (1977a). A review of criminality among twins. In S.A. Mednick & K.O. Christiansen (Eds.), *Biosocial bases of criminal behavior* (pp. 45-88). New York: Gardner Press.

Christiansen, K.O. (1977b). A preliminary study of criminality among twins. In S.A. Mednick & K.O. Christiansen (Eds.), *Biosocial bases of criminal behavior* (pp. 89-108). New York: Gardner Press.

Cloninger, C.R., Christiansen, K.O., Reich, T., & Gottesman, I.I. (1978). Implications of sex differences in the prevalences of antisocial personality, alcoholism, and criminality for familial transmission. *Archives of General Psychiatry,* **35**, 941-951.

Cloninger, C.R., & Gottesman, I.I. (1987). Genetic and environmental factors in antisocial behavior disorders. In S.A. Mednick, T.E. Moffitt, & S. Stack (Eds.), *The causes of crime: New biological approaches* (pp. 92-109). Cambridge: Cambridge University Press.

Cloninger, C.R., Sigvardsson, S., Bohman, M., & von Knorring, A.L. (1982). Predisposition to petty criminality in Swedish adoptees: II. Cross-fostering analysis of gene-environmental interactions. *Archives of General Psychiatry,* **39**, 1242-1247.

Coccaro, E.F., Bergeman, C.S., Kavoussi, R.J., & Seroczynski, A.D. (1997). Heritability of aggression and irritability: A twin study of the Buss-Durkee Aggression Scales in adult male subjects. *Biological Psychiatry,* **41**, 273-284.

Coccaro, E.F., & Kavoussi, R.J. (1996). Neurotransmitter correlates of impulsive aggression. In D.M. Stoff & R.B. Cairns (Eds.), *Aggression and violence: Genetic, neurobiological, and biosocial perspectives* (pp. 67-99). Mahwah: Erlbaum Associates.

Cohen, J. (1988). *Statistical power analysis for the behavioral sciences* (2nd ed.). Hillsdale, NJ: Erlbaum.

Coid, B., Lewis, S.W., & Reveley, A.M. (1993). A twin study of psychosis and criminality. *British Journal of Psychiatry,* **162**, 87-92.

Comings, D.E. (2000). The role of genetics in ADHD and conduct disorder – Relevance to the treatment of recidivistic antisocial behavior. In D.H. Fishbein (Ed.), *The science, treatment, and prevention of antisocial behaviors* (pp. 16-1 – 16-26). Kingston, NJ: Civic Research Institute, Inc.

Crelia, R.A., & Tesser, A. (1996). Attitude heritability and attitude reinforcement: A replication. *Personality and Individual Differences,* **21**, 803-808.

Crowe, R. (1974). An adoption study of antisocial personality. *Archives of General Psychiatry,* **31**, 785-791.

Dalgaard, O.S., & Kringlen, E. (1976). A Norwegian twin study of criminality. *British Journal of Clinical Psychology,* **16**, 213-232.

Eley, T.C., Lichtenstein, P., & Stevenson, J. (1999). Sex differences in the etiology of aggressive and nonaggressive antisocial behavior: Results from two twin studies. *Child Development,* **70**, 155-168.

Elliott, F.A. (1988). Neurological factors. In V.B. Van Hassett, R.L. Morrison, A.S. Bellack, & M. Hersen (Eds.), *Handbook of family violence* (pp. 359-382). New York: Plenum.

Ellis, L., & Walsh, A. (2000). *Criminology: A global perspective.* Boston, MA: Allyn & Bacon.

Eysenck, H.J. (1977). *Crime and personality* (3rd ed.). St. Albans: Paladin.

Eysenck, H.J. (1997). Personality and the biosocial model of antisocial and criminal behavior. In A. Raine, P. Brennan, D.P. Farrington, & S.A. Mednick (Eds.), *Biosocial bases of violence* (pp. 21-38). New York: Plenum.

Falconer, D.S. (1965). The inheritance of liability to certain diseases, estimated from the incidence among relatives. *Annals of Human Genetics*, **29**, 51-76.

Fava, M., Vuolo, R.D., Wright, E.C., Nierenberg, A.A., Alpert, J.E., & Rosenbaum, J.F. (2000). Fenfluramine challenge in unipolar depression with and without anger attacks. *Psychiatry Research*, **94**, 9-18.

Ge, X., Cadoret, R.J., Conger, R.D., Neiderhiser, J.M., Yates, W., Troughton, E., & Stewart, M.A. (1996). The developmental interface between nature and nurture: A mutual influence model of child antisocial behavior and parent behaviors. *Developmental Psychology*, **32**, 574-589.

Ghodsian-Carpey, J., & Baker, L.A. (1987). Genetic and environmental influences on aggression in 4- to 7-year old twins. *Aggressive Behavior*, **13**, 173-186.

Goldman, D., & Fishbein, D.H. (2000). Genetic bases for impulsive and antisocial behaviors: Can their course be altered? In D.H. Fishbein (Ed.), *The science, treatment, and prevention of antisocial behaviors* (pp. 9-1 – 9-18). Kingston, NJ: Civic Research Institute, Inc.

Gottfredson, M.R., & Hirschi, T. (1990). *A general theory of crime*. Stanford: Stanford University Press.

Grove, W.M., Eckert, E.D., Heston, L., Bouchard, T.J., Segal, N., & Lykken, D.T. (1990). Heritability of substance abuse and antisocial behavior: A study of monozygotic twins reared apart. *Biological Psychiatry*, **27**, 1293-1304.

Hibbeln, J.R., Umhau, J.C., Linnoila, M., George, D.T., Ragan, P.W., Shhoaf, S.E., Vaughan, M.R., Rawlings, R., & Salem, N. (1998). A replication study of violent and nonviolent subjects: cerebrospinal fluid metabolites of serotonin and dopamine are predicted by plasma essential fatty acids. *Biological Psychiatry*, **44**, 243-249.

Higley, J.D., King, S.T., Hasert, M.F., Champoux, M., Suomi, S.J., & Linnoila, M. (1996). Stability of interindividual differences in serotonin function and its relationship to severe aggression and competent social behavior in rhesus macaque females. *Neuropsychopharmacology*, **14**, 67-76.

Higley, J.D., Mehlman, P.T., Higley, S.B., Fernald, B., Vickers, J., Lindell, S.G., Taub, D.M., Suomi, S.J., & Linnoila, M. (1996). Excessive mortality in young free-ranging male nonhuman primates with low cerebrospinal fluid 5-Hydroxyindoleacetic acid concentrations. *Archives of General Psychiatry*, **53**, 37-543.

Hudziak, J.J., Rudiger, L.P., Neale, M.C., Heath, A.C., & Todd, R.D. (2000). A twin study of inattentive, aggressive, and anxious/depressed behaviors. *Journal of the American Academy of Child and Adolescent Psychiatry*, **39**, 469-476.

Jarey, M.L., & Stewart, M.A. (1985). Psychiatric disorder in the parents of adopted children with aggressive conduct disorder. *Neuropsychobiology*, **13**, 7-11.

Kendler, K.S. (1983). Twin studies of schizophrenia: a current perspective. *American Journal of Psychiatry,* **140**, 1143-1425.

Kuikka, J.T., Tiihonen, J., Bergstrom, K.A., Karhu, J., Rasanen, P., & Eronen, M. (1998). Abnormal structure of human striatal dopamine re-uptake sites in habitually violent alcoholic offenders: a fractal analysis. *Neuroscience Letters,* **253**, 195-1973.

Kuruoglu, A.C., Arikan, Z., Vural, G., Karatas, M., Arak, M., & Isik, E. (1996). Single photon emission computerised tomography in chronic alcoholism: Antisocial personality disorder may be associated with decreased frontal perfusion. *British Journal of Psychiatry,* **169**, 348-354.

Linnoila, V.M., & Virkkunen, M. (1992). Aggression, suicidality, and serotonin. *Journal of Clinical Psychology,* **53**, 46-51

Linnoila, M., Virkkunen, M., Scheinin, M., Nuutila, A., Rimon, R., & Goodwin, F.K. (1983). Low cerebrospinal fluid 5 hydroxyindoleacetic acid concentration differentiates impulsive from nonimpulsive violent behavior. *Life Sciences,* **33**, 2609-2614.

Lyons, M.J. (1996). A twin study of self-reported criminal behavior. In G.R. Bock & J.A. Goode (Eds.), *Genetics of criminal and antisocial behavior* (pp. 61-75). West Sussex, England: John Wiley and Sons.

Lytton, H., Watts, D., & Dunn, B.E. (1988). Stability of genetic determination from age 2 to age 9: A longitudinal twin study. *Social Biology,* **35**, 62-73.

Mednick, S.A., & Finello, K.M. (1983). Biological factors and crime: Implications for forensic psychiatry. *International Journal of Law and Psychiatry,* **6**, 1-15.

Mednick, S.A., Gabrielli, W.H., & Hutchings, B. (1984). Genetic influences in criminal convictions: Evidence from an adoption cohort. *Science,* **224**, 891-894.

Meyer, J.M., Rutter, M., Silberg, J.L., Maes, H.H., Simonoff, E., Shillady, L.L., Pickles, A., Hewitt, J.K., & Eaves, L.J. (2000). Familial aggregation for conduct disorder symptomatology: the role of genes, marital discord, and family adaptability. *Psychological Medicine,* **30**, 759-774.

Miczek, K.A., Haney, M., Tidey, J., Vivian, J., & Weerts, E. (1994). Neurochemistry and pharmacotherapeutic management of aggression and violence. In A.J. Reiss, K.A. Miczek, & J.A. Roth (Eds.), *Understanding and preventing violence: Volume 2, Biobehavioral influences* (pp. 245-514). Washington, DC: National Academy Press.

Miczek, K.A., Weerts, E., Haney, M., & Tidey, J.W. (1994). Neurobiological mechanisms controlling aggression: preclinical developments for pharmacotherapeutic interventions. *Neuroscience & Biobehavioral Reviews,* **18**, 97-110.

Miles, D.R., & Carey, G. (1997). Genetic and environmental architecture of human aggression. *Journal of Personality and Social Psychology,* **72**, 207-217.

Moffitt, T.E. (1987). Parental mental disorder and offspring criminal behavior: An adoption study. *Psychiatry: Interpersonal and Biological Processes,* **50**, 346-360.

Moffitt, T.E. (1993). The neuropsychology of conduct disorder. *Development and Psychopathology*, **5**, 135-151.

Moilanen, I. (1987). Dominance and submissiveness between twins. *Acta Geneticae Medicae et Gemellologiae*, **36**, 249-255.

Moore, T.M., Scarpa, A, & Raine, A. (manuscript submitted for review). A meta-analysis of serotonin 5-HIAA and antisocial behavior.

O'Connor, M., Foch, T., Sherry, T., & Plomin, R. (1980). A twin study of specific behavioral problems of socialization as viewed by parents. *Journal of Abnormal Child Psychology*, **8**, 189-199.

O'Connor, T.G., Deater-Deckard, K., Fulker, D., Rutter, M., & Plomin, R. (1998). Genotype-environment correlations in late childhood and early adolescence: antisocial behavioral problems and coercive parenting. *Developmental Psychology*, **34**, 970-981.

O'Keane, V., Moloney, E., O'Neill, H., O'Connor, A., Smith, C., & Dinan, T.G. (1992). Blunted prolactin responses to d-Fenfluramine in sociopathy: Evidence for subsensitivity of central serotonergic function. *British Journal of Psychiatry*, **160**, 643-646.

Plomin, R., Owen, M.J., & McGuffin, P. (1994). The genetic basis of complex human behaviors. *Science*, **264**, 1733-1739.

Raine, A. (1993). *The psychopathology of crime: Criminal behavior as a clinical disorder*. San Diego: Academic Press.

Raine, A. (1997). *Psychopathology and tonic heart rate*. Paper presented at the Society for Research in Psychopathology, Palm Springs, October 23-26.

Raine, A., Lencz, T., Bihrle, S., LaCasse, L., & Colletti, P. (2000). Reduced prefrontal gray matter volume and reduced autonomic activity in antisocial personality disorder. *Archives of General Psychiatry*, **57**, 119-127.

Raine, A., Meloy, J.R., Bihrle, S., Stoddard, J., LaCasse, L., & Buchsbaum, M.S. (1998). Reduced prefrontal and increased subcortical brain functioning assessed using positron emission tomography in predatory and affective murderers. *Behavioral Sciences & the Law*, **16**, 319-332.

Raine, A., Venables, P.H., & Williams, M. (1990). Relationships between CNS and ANS measures of arousal at age 15 and criminality at age 24. *Archives of General Psychiatry*, **47**, 1003-1007.

Rowe, D.C. (1983). Biometric genetic models of self-reported delinquent behavior: A twin study. *Behavior Genetics*, **13**, 473-489.

Rushton, J.P., Fulker, D.W., Neale, M.C., Nias, D.K.B., & Eysenck, H.J. (1986). Altruism and aggression: The heritability of individual differences. *Journal of Personality and Social Psychology*, **50**, 1192-1198.

Rutter, M. (1996). Introduction: concepts of antisocial behavior, of cause, and of genetic influences. In G.R. Bock & J.A. Goode (Eds.), *Genetics of criminal and antisocial behavior* (pp. 1-15). Chichester: Wiley.

Rutter, M., Maughan, B., Meyer, J., Pickles, A., Silberg, J., Simonoff, E., & Taylor, E. (1997). Heterogeneity of antisocial behavior: Causes, continuities, and consequences. *Nebraska Symposium on Motivation*, **44**, 45-118.

Rutter, M., Boltin, P., Harrington, R., Le Couteur, A., Macdonald, H., & Simonoff, E. (1990). Genetic factors in child psychiatric disorders I. A review of research strategies. *Journal of Child Psychology and Psychiatry*, **31**, 5-37.

Rutter, M., Silberg, J., O'Connor, T., & Simonoff, E. (1999a). Genetics and child psychiatry: I. Advances in quantitative and molecular genetics. *Journal of Child Psychology and Psychiatry*, **40**, 3-18.

Rutter, M., Silberg, J., O'Connor, T., & Simonoff, E. (1999b). Genetics and child psychiatry: II. Empirical research findings. *Journal of Child Psychology and Psychiatry*, **40**, 19-55.

Rydenour, T.A. (2000). Genetic epidemiology of antisocial behavior. In D.H. Fishbein (Ed.), *The science, treatment, and prevention of antisocial behaviors* (pp. 7-1 – 7-24). Kingston, NJ: Civic Research Institute, Inc.

Scarr, S., & McCartney, K. (1983). How people make their own environments: A theory of genotype—environment effects. *Child Development*, **54**, 424-435.

Schacter, F.F., & Stone, R.K. (1985). Difficult sibling, easy sibling: temperament and within family environment. *Child Development*, **54**, 424-435.

Schwesinger, G. (1952). The effect of differential parent-child relations on identical twin resemblance in personality. *Acta Geneticae Medicae et Gemellologiae*, **1**, 40-47.

Sheley, J.F. (1991). *Criminology: A contemporary handbook.* Belmont, CA: Wadsworth Publishing, Co.

Shields, J. (1977). Polygenic influences. In M. Rutter & L. Hersov (Eds.), *Child psychiatry: Modern approaches* (pp. 22-46). Oxford: Blackwell.

Sigvardsson, S., Cloninger, R., Bohman, M., & von Knorring, A.L. (1982). Predisposition to petty criminality in Swedish adoptees. III. Sex differences and validation of the male typology. *Archives of General Psychiatry*, **39**, 1248-1253.

Silverton, L. (1988). Crime and the schizophrenia spectrum: A study of three Danish cohorts. In T.E. Moffitt & S.A. Mednick (Eds.), *Biological contributions to crime causation* (pp. 93-120). Dordrecht: Martinus Nijhoff.

Simonoff, E., Pickles, A., Meyer, J., Silberg, J., & Maes, H. (1998). Genetic and environmental influences on subtypes of conduct disorder behavior in boys. *Journal of Abnormal Child Psychology*, **26**, 495-509.

Slater, E. (1953). The incidence of mental disorder. *Annals of Eugenics*, **6**, 172.

Slutske, W.S., Heath, A.C., Dinwiddie, S.H., Madden, P.A.F., Bucholz, K.K., Dunne, M.P., Statham, D.J., & Martin, N.G. (1997). Modeling genetic and environmental influences in the etiology of conduct disorder: A study of 2682 twin pairs. *Journal of Abnormal Psychology*, **106**, 266-279.

Stanley, B., Molcho, A., Stanley, M., Winchel, R., Gameroff, M.J., Parsons, B., & Mann, J.J. (2000). Association of aggressive behavior with altered serotonergic function in patients who are not suicidal. *American Journal of Psychiatry*, **157**, 609-614.

Stevenson, J., & Graham, P. (1988). Behavioral deviance in 13-year-old twins: An item analysis. *Journal of the American Academy of Child and Adolescent Psychiatry*, **27**, 791-797.

Suomi, S.J. (2000). A biobehavioral perspective on developmental psychopathology: Excessive aggression and serotonergic dysfunction in monkeys. In A.J. Sameroff, M. Lewis, & S.M. Miller (Eds.), *Handbook of developmental psychopathology* (2nd ed., pp. 237-256). New York: Kluwer Academic/ Plenum Publishers.

Susman, E.J., Dorn, L.D., & Chrousos, G.P. (1991). Negative affect and hormone levels in young adolescents: Concurrent and predictive perspectives. *Journal of Youth and Adolescence*, **20**, 167-190.

Sweet, R.A., Nimgaonkar, V.L., Vishwajit, L., Kamboh, M.I., Lopez, O.L., Zhang, F., & DeKosky, S.T. (1998). Dopamine receptor genetic variation, psychosis, and aggression in Alzheimer disease. *Archives of Neurology*, **55**, 1335-1340.

Tiihonen, J., Kuikka, J., Bergstrom, K., Hakola, P., Karhu, J., Ryynanen, O.P., & Fohr, J. (1995). Altered striatal dopamine reuptake site densities in habitually violent and non-violent alcoholics. *Nature Medicine*, **1**, 645-657.

Trasler, G. (1987). Biogenetic factors. In H.C. Quay (Ed.), *Handbook of juvenile delinquency* (pp. 184-216). New York: Wiley.

Unis, A.S., Cook, E.H., Vincent, J.G., Gjerde, D.K., Perry, B.D., Mason, C., & Mitchell, J. (1997). Platelet serotonin measures in adolescents with conduct disorder. *Biological Psychiatry*, **42**, 553-559.

Venables, P.H., & Raine, A. (1987). Biological theory. In B. McGurk, D. Thornton, & M. Williams (Eds.), *Applying psychology to imprisonment: Theory and practice* (pp. 3-28). London: HMSO.

Vernon, P.A., McCarthy, J.M., Johnson, A.M., Jang, K.L., & Harris, J.A. (1999). Individual differences in multiple dimensions of aggression: a univariate and multivariate genetic analysis. *Twin Research*, **2**, 16-21.

Virkkunen, M., Rawlings, R., Tokola, R., Poland, R.E., Guidotti, A., Nemeroff, C., Bissette, G., Kalogeras, K., Karonen, S.L., & Linnoila, M. (1994). CSF biochemistries, glucose metabolism, and diurnal activity rhythms in alcoholic, violent offenders, fire setters, and healthy volunteers. *Archives of General Psychiatry*, **51**, 20-27.

Wadsworth, M.E.J. (1976). Delinquency, pulse rate and early emotional deprivation. *British Journal of Criminology*, **16**, 245-256.

Wahlsten, D. (1990). Insensitivity of the analysis of variance to heredity-environment interaction. *Behavioral and Brain Sciences*, **13**, 109-161.

Wilson, J.Q., & Herrnstein, R. (1985). *Crime and human nature*. New York: Simon & Schuster.

Wilson, R.S. (1983). The Louisville Twin Study: Developmental asynchronies in behavior. *Child Development*, **20**, 203-316.

Chapter 5

COGNITIVE ANTECEDENTS OF VIOLENCE AND AGGRESSION

Adrienne K. Elliott, M.A. and Allan F. Mirsky, Ph.D.
Section on Clinical and Experimental Neuropsychology, NIMH/NIH

INTRODUCTION

Research on the antecedents of aggression and violence has increased considerably in recent years, as parents, researchers, and policy makers endeavor to understand this phenomenon. Many theories have been proposed (e.g., Pincus, 2001) and numerous intervention strategies have been tried (see Elliott & Tolan 1999), with varying degrees of success. Many factors contribute to the development of aggressive behavior, and although research offers some general predictors, there remains a large degree of uncertainty that at present can not be captured by a single theory. In this chapter, we explore the role of impaired or altered cognition in the genesis of aggressive behavior: what neuropsychological deficits are observed in aggressive individuals, what might be contributing to or causing these deficits, and how do these limitations influence behavior? Special emphasis is given to the frontal lobes and executive functioning, as well as putative brain disease and a number of environmental factors that contribute directly and indirectly to aggression. While impaired cognition may not account for all the variance, it seems to represent an important piece of the aggression puzzle. Specifically, it is hypothesized that cognitive deficits, especially impulsivity, poor planning ability, mental inflexibility, low verbal intelligence, and impaired attention, limit an individual's ability to cope with other biological and environmental vulnerabilities; these limitations, in turn, lead to feelings of frustration and anxiety and ultimately, to difficulty with regulation of emotion and increased aggressive behavior. Some consideration will be given, as well, to the notion of resilience and other factors that might protect an at-risk individual from developing behavior problems, such as aggression.

Aggressive individuals have a multitude of cognitive deficits, which are mostly attributed to damaged frontal lobes and impaired executive functioning, including impulsivity, poor planning ability, working memory deficits, and low verbal intelligence (see Filley *et al.*, 2001; Séguin, Pihl, Harden, Tremblay, & Boulerice, 1995). Although aggression is not necessarily seen in patients with attention deficit hyperactivity disorder (ADHD), aggressive behavior is more common in children who have both ADHD and conduct disorder (Pliszka, 1998). Considerable attention has been given to the influence of putative brain disease, including epilepsy, schizophrenia, and alcoholism. Although some researchers have speculated about a higher incidence of aggression in epileptic patients, this does not appear to be true (Steinert & Froscher, 1994). Similarly, although many people believe that schizophrenic patients are violent, evidence suggests that this is true only when other factors are present, such as substance abuse and medication noncompliance (Spearing, 1999). A higher incidence of aggressive behavior is observed in offspring who were exposed prenatally to alcohol (Olson *et al.*, 1997) and there is a link between violent criminal behavior and substance use (Department of Justice, 1996). There are also a number of environmental factors that have been linked to aggressive behavior, including having a rejecting mother and a criminal parent (Raine, Brennan, Mednick & Mednick, 1996), low socioeconomic status (Gartner & Whitaker-Azmitia, 1996), a history of physical and sexual abuse (Shields & Cicchetti, 1998), inadequate nutrition (Kleinman *et al.*, 1998), and exposure to environmental neurotoxins, such as lead (Needleman, Riess, Tobin, Biesecker, & Greenhouse, 1996). Finally, additional variables that appear to affect aggression, which will only be discussed briefly, include neurobiological factors (e.g., Adamec, 1991) and genetic contributions (e.g., Alsobrook & Pauls, 2000). Aggressive behavior involves a combination of factors, including cognitive deficits which may limit an individual's ability to manage other biological and environmental vulnerabilities.

Definitions and Methodology

One difficulty in reviewing the literature in this area is the diversity of definitions and terminology used to represent the topic. Although there is inconsistency in the literature, for the purpose of this chapter the term *aggression* is used broadly and is meant to encompass behavior that infringes upon the rights of other people, including violence and noncriminal aggression. Self-injurious behavior, including suicide, and institutional aggression are not addressed here (for a review of that literature, see Chapter 9). Descriptions of participant samples are included to clarify results that are specific to a particular population or category of aggression. Also, many of the studies reviewed here suffer from sampling bias and are not epidemiological; thus their generalizability is limited. It is our hope that the reader will gain some understanding of several important variables in the area while also recognizing that the state of the art is limited. Finally, although many studies were reviewed for this chapter, it represents a small sampling of a large literature, meaning that the information presented should be used as a preliminary resource for more in-depth analyses.

COGNITIVE FUNCTIONING

Many aggressive or violent individuals have cognitive deficits as well as impaired social cognition (Harvey, Fletcher, & French, 2001). Beyond the immediate effects of these deficits, cognitive impairments have broader behavioral and psychological implications, including increased emotional lability due to frustration or anxiety, impulsivity, and impaired judgement. While exploring the relationship between cognition and aggression is of theoretical and practical importance, a bit of caution is warranted. Associating cognitive functions with specific brain structures or neuronal pathways requires a degree of inference. Therefore, assertions about a link between biological anomalies associated with some cognitive deficit and aggressive behavior must be made still more tentatively. Nonetheless, studies of cognition and aggression are relevant to the extent that impairment in cognitive processes is a sign of cerebral compromise. Additionally, it is important to distinguish between causation and association. Although certain cognitive deficits may be causally related to a particular area of the brain, observing these deficits in an aggressive individual does not necessarily mean that the compromised area is causing the offensive behavior.

A Surfeit of Deficits

Neuropsychological research suggests that frontal lobe dysfunction, particularly deficits in executive functioning, is most likely to be associated with aggressive behavior (see Filley *et al.*, 2001; Séguin *et al.*, 1995). There are monosynaptic connections between the brain stem and the frontal lobe. Thus, what appears to be frontal lobe compromise may actually be a sign of brain stem injury (Goldberg, Bilder, Mattis, Antin, & Hughes, 1989). Aggressive individuals show more signs of impulsivity on neuropsychological measures than do nonaggressive individuals (e.g., Stanford, Greve, & Gerstle, 1997). Aggressive boys have working memory difficulties, particularly when organizing increasing amounts of information (Séguin, Boulerice, Harden, Tremblay, & Pihl, 1999). In a study of inner-city boys and girls, working memory deficits and impulsivity both predicted teacher-rated aggression in the classroom (Sun, 2001). Although it is not exclusively an executive function, aggressive individuals and violent offenders also seem to have lower verbal intelligence (see Bassarath, 2001; Déry, Toupin, Mercier, & Fortin, 1999; Farrington, & Loeber, 2000; Lyman, Moffitt, & Stouthamer-Loeber, 1993; Moffitt, 1990; Pineda *et al.*, 2000; Teichner & Golden, 2000). There has also been speculation about the relationship between reading disorder and aggression. A review of the literature suggests that there is not enough evidence to substantiate a causal relationship between reading disorder and aggression (Cornwall & Bawden, 1992), although these two difficulties may nonetheless be associated (see Spreen, Risser, & Edgell, 1995). Overall, aggressive adults show several signs of frontal-lobe deficits, such as impulsivity, cognitive inflexibility, and poor planning ability (see Golden, Jackson, Peterson-Rohne, & Gontkovsky, 1996). In aggressive adolescents, frontal-lobe dysfunction is also evident, with deficits in concept formation, abstract reasoning, mental flexibility, planning, and goal setting (Teichner & Golden, 2000). Additional coexisting problems may include learning disabilities, such as reading disorder, and lower verbal intelligence.

Impulsivity and Attention

Impulsivity and impaired attention have been studied extensively in many populations, including aggressive individuals. In a prospective study of males and females, Babinski, Hartsough, and Lambert (1999) found that hyperactivity/impulsivity and early conduct problems predict later criminal behavior but inattention does not. In another study, aggressive college students showed poorer performance on several common neuropsychological measures (Trails B, the Wisconsin Card Sorting Task, and the Controlled Oral Word Association Test; Stanford *et al.*, 1997), with deficits in impulse control and verbal strategic processing. Although this result points to a problem with impulsivity, all of the neuropsychological measures used in Stanford *et al.*'s work measure attention as well, suggesting the possibility of difficulties in that domain. Finally, in a study of attention in violent offenders (Bergvall, Wessely, Forsman, & Hansen, 2001), participants had difficulty shifting mentally from one perceptual category to another.

Given the association between impulsivity and aggression, studies of the possible relationship between ADHD and aggression have been conducted, yielding mixed results. Some of the confusion regarding the degree to which ADHD is related to aggression stems from the co-occurrence of other conditions, such as Oppositional Defiant Disorder and Conduct Disorder. In a review, Pliszka (1998) concludes that ADHD is a heterogeneous disorder and that ADHD with conduct disorder may represent a distinct subtype. He notes further that children who are diagnosed with ADHD, conduct disorder, and reading disabilities are especially prone to antisocial behavior later in life. While it may be difficult to sort out the effect of ADHD from that of other conditions, the reduced inhibition associated with ADHD may extend to more general difficulties such as regulating aggression, frustration, and mood (Baird, Stevenson, & Williams, 2000). While the impulsivity seen in children with ADHD may lead to difficulties with emotion regulation, this does not mean that all or even most children with ADHD will go on to commit violent acts. Additionally, while some aggressive persons with ADHD have multiple diagnoses and while some may experience difficulties regulating emotions, ADHD does not account for the association between impaired executive functioning and aggression (Séguin *et al.*, 1999).

PUTATIVE BRAIN DISEASE

While uncovering the precise combination of variables that are predictive of aggression remains elusive, one area that has received some attention is that of putative brain disease, or disorders that are presumed to be caused by a neurological disturbance. Although aggressive behavior is not always seen in neurologic and psychiatric conditions, many of these disorders are associated with cognitive deficits. While cognitive impairments are not the sole cause of aggressive behavior, they may increase feelings of frustration as well as limit a person's ability to manage other serious vulnerabilities. The resulting emotional dysregulation, in conjunction with impulsivity and impaired judgement, may lead an individual to act aggressively. While psychiatric and neurological disorders may make a particular individual more vulnerable, these conditions in and of themselves do not sufficiently account for the emergence of aggressive behavior, even when cognitive deficits are present.

Epilepsy

There has been a great deal of debate over the years concerning the relationship between epilepsy and aggression. Although they have been labeled by some as aggressive, the jerks and movements exhibited during a seizure (ictal) are symptomatic rather than intentional. Further, although there has been some speculation about a higher incidence of aggressive behavior in epileptic patients between seizures (interictal), this is not substantiated in the majority of cases. Although aggression is not typical of epileptic patients, like any neurologic or psychiatric group, some persons with seizure disorders do engage in aggressive behavior. When aggression is prominent so too are other limiting factors.

Several studies have explored factors that may increase the likelihood of aggressive acts in epileptic patients. A retrospective study of temporal lobe patients indicates that aggressive individuals are more likely to be male and to have a lower mean intelligence, employment difficulties, early onset of seizure disorder, and a history of behavior problems (Herzberg & Fenwick, 1988). In a study comparing epileptic patients with or without other psychiatric difficulties, Steinert and Froscher (1994) found that there was more evidence of aggressive behavior in the psychiatric group, and concluded that aggressive behavior seen in epileptic patients is tied to the presence of other psychiatric disorders and not to the timing of the seizures. Tebartz van Elst, Woermann, Limieux, Thompson, and Trimble (2000) studied epileptic patients (temporal lobe) with or without intermittent explosive disorder (IED) and concluded that IED is associated with left-side or bilateral EEG abnormalities, low intelligence, and depression and anxiety. Finally, in a study comparing epileptic patients with or without a history of violence, violent individuals had more schizophrenia spectrum disorders and more mental retardation than did nonviolent participants (Mendez, Doss, & Taylor, 1993). Although both cognitive deficits and epilepsy presumably stem from brain disease, Mendez *et al.*'s study suggests that aggressive behavior may be related to psychopathology and cognitive deficits rather than to epilepsy.

In a study comparing the personalities of epileptic patients with mild to moderate cognitive impairments and a matched control group comprised of non-epileptic patients, there appeared to be no statistically significant personality differences between the two groups, nor did the epileptic individuals have a more aggressive personality style (Deb & Hunter, 1991). While there was some evidence of personality pathology in both groups, the severity of the disturbance did not differ across groups, again suggesting that aggression is not necessarily an associated feature of epilepsy. In a study comparing nonaggressive epileptic patients and controls to patients with temporal-lobe epilepsy with a history of aggression, the aggressive group had a decrease in gray matter, mostly in the left frontal lobe (Tebartz van Elst *et al.*, 2000). Finally, in a study of aggressive children with intractable partial epilepsy, all had developmental delays and showed signs of prefrontal glucose hypometabolism as compared to normal adults and nonaggressive children with epilepsy (Juhász, Behen, Muzik, Chugani, & Chugani, 2001). It seems, therefore, that a neurological disorder other than that leading to seizures may be underlying the aggressive behavior seen in some patients.

Some researchers have speculated that aggression might be associated with particular seizure types rather than with seizures in general. Although they agree that aggressive behavior is not common in epileptic patients, Devinsky *et al.* (1994) note that patients with temporal-lobe epilepsy are a heterogeneous group, and that the degree to which individuals report hostile feelings and behaviors varies by seizure type. Although Devinsky *et al.* looked at very specific seizure types, the authors of an earlier study suggest that type of epilepsy, either temporal-lobe or generalized, is not related to aggression (Hermann, Schwartz, Whitman, & Karnes, 1980). Steinert and Froscher (1994) concluded similarly that aggressive behavior is not related to seizure type. While it is possible that having a specific type of seizure disorder serves as a vulnerability for later problems, the behavior seen in aggressive individuals appears to be related to social, cognitive, and neurologic difficulties other than epilepsy.

Schizophrenia

There exists a pervasive assumption regarding the frequency of aggressive behavior in persons with psychotic disorders, including schizophrenia. According to the Department of Health and Human Services (2001), 2.2 million American adults suffer from schizophrenia, or 1.1% of the population. Of those persons with the disorder, the vast majority do not commit violent crimes. If violence does occur, it is usually associated with a prior history of violent behavior, substance abuse, medication noncompliance, and paranoid ideation (Spearing, 1999). Also, if a schizophrenic person becomes violent his or her aggression typically occurs in the home and is directed toward a friend or family member rather than a stranger (Spearing, 1999). Krakowski, Volavka, and Brizer (1986) point out that the disorganization seen in psychotic states might preclude the occurrence of aggressive behavior. Robertson (1988) notes, similarly, that because their odd behavior makes them more noticeable, schizophrenic patients' aggressive transgressions are more likely to be detected; this may also contribute to the perception that these individuals are violent. The disorganization, unusual thinking, and impaired judgement seen in schizophrenic patients strongly suggest a cognitive component to this disorder (see Frith, 2000). Overall, these patients seem to have substantial neuropsychological compromise independent of intelligence (Kremen, Seidman, Faraone, & Tsuang, 2001), including impaired attention (Mirsky, Yardley, Jones, Walsh, & Kendler, 1995) and working memory deficits (Peled *et al.*, 2001). As Krakowski *et al.* (1986) note, extensive impairment might actually thwart aggressive action while more modest levels of pathology and cognitive impairment might increase the potential for aggressive behavior.

In a 5-year prospective study, inpatients with schizophrenia or other delusional disorders were found to be more likely to commit violent acts than were patients with other diagnoses, although the overall prevalence of violence was low. Of those psychiatric patients who became aggressive, most also had a history of violence and previous psychiatric admissions (Grassi, Peron, Marangoni, Zanchi, & Vanni, 2001). Past violent behavior is known to be a predictor of future aggressive acts independent of schizophrenia, suggesting that the psychiatric diagnosis may play only a modest role. Because inpatient samples and patients with a criminal history are not representative of the larger population of schizophrenic patients, no firm conclusions can be drawn about community patients or persons with no criminal history. Furthermore, for unmedicated schizophrenic patients, admittance to a psychiatric facility may exacerbate paranoia and delusions, possibly resulting in more acting-out behavior, especially during the early days of treatment. This possibility is further supported by the fact that the frequency of violent behavior decreases after neuroleptics are introduced (Steinert, Sippach, & Gebhardt, 2000). Based on the evidence reviewed here, schizophrenia and its associated cognitive deficits serve as a vulnerability to aggression, especially in the presence of other factors such as medication noncompliance, substance abuse, and a prior history of violent behavior.

Alcoholism

There are at least two identifiable points at which alcohol might contribute to cognitive compromise and aggressive behavior. Prenatal exposure to alcohol is known to have a number of effects on the fetus. Additionally, intoxication or prolonged alcohol abuse are also closely associated with reduced inhibition and sometimes, aggressive behavior. In the case of prenatal exposure, Streissguth and her research group have done extensive work in the area as part of the Seattle Longitudinal Study on Alcohol and Pregnancy. Initiated in 1974, this project involves a large scale prospective study of the offspring of approximately 500 women who reported using alcohol during pregnancy (see Streissguth, Martin, Martin, & Barr, 1981). Fetal Alcohol Syndrome (FAS) and Fetal Alcohol Effect (FAE) are associated with a number of deficits, including cognitive, social, legal, and academic problems. Individuals with FAS or FAE have lower intelligence and show poorer performance on measures of reading, spelling, and arithmetic (Streissguth, Barr, Kogan, & Bookstein, 1997). Additionally, of the 661 offspring studied, all of whom had been exposed prenatally to alcohol, many had experienced legal, employment, and academic difficulties and had been physically or sexually abused. In addition to processing-speed difficulties and some memory problems (Olson, Feldman, Streissguth, Sampson, & Bookstein, 1998), adolescents with FAS and FAE often have attentional difficulties. Specifically, these children have difficulty focusing and shifting their attention, as well as exhibiting impairment on tests of sustained attention or vigilance (Connor, Streissguth, Sampson, Bookstein, & Barr, 1999). The deficit is most pronounced on the auditory Continuous Performance Test (CPT) and is seen to a lesser extent on visual CPT tasks. Higher verbal intelligence to some extent is associated with better performance on the CPT (Pascualvaca *et al.*, 1997), suggesting that auditory processing deficits and impaired attention in these children might feed into frustration, and ultimately lead to aggression. Overall, adolescents with FAS show a cluster of problems including antisocial behavior, substance abuse, school difficulties, negative self-esteem, impulsivity, and disorganization (Olson *et al.*, 1997), all of which may contribute to aggression.

While individuals with FAS and FAE experience a number of difficulties, some who were exposed prenatally to alcohol thrive with relatively few difficulties. Several protective factors have been identified that may account for this disparity. In a study of individuals with FAS and FAE, Streissguth *et al.* (1997) identified eight protective factors that seem to diminish the emergence of significant difficulties. They include living in a stable environment for 72% of one's life, being diagnosed with FAS or FAE before age 6, having never been the victim of violence, remaining in each living situation for approximately 3 or more years, experiencing a quality home environment from age 8 to 12, being found eligible for special services, having a diagnosis of FAS rather than FAE, and having basic needs met for at least 13% of one's life. Streissguth *et al.*'s research provides a basis for hypothesizing about why certain individuals with FAS or other neurologic or psychiatric conditions do not go on to lead troubled lives. Future research that compares the family, medical, social, and occupational histories of aggressive and nonaggressive individuals might serve to enhance our understanding of why some people thrive in spite of serious vulnerabilities.

In addition to prenatal exposure, alcohol abuse can also significantly impact one's life, resulting in numerous problems, among them increased levels of aggression and cognitive compromise. A 1996 study of jailed inmates by the Department of Justice reveals that approximately 55% of violent offenders were on drugs, alcohol, or both at the time of their offense. Of violent offenders, approximately 28% had been drinking prior to perpetrating their crime. A study of Detroit census data shows that rates of crime increase with the availability of alcohol, suggesting that reducing availability might curb criminal behavior (Gyimah-Brempong, 2001). A study of nonhuman primates indicates that individual differences in neurotransmitter levels predict alcohol-induced aggression, as well as aggression and impulsivity more generally (Higley, 2001). Differences in neurotransmitter balance may thus relate to individual propensity to become violent when intoxicated. Although studies linking psychiatric disorders to aggression have been equivocal, in the case of people with antisocial personality disorder (ASPD), there is evidence of higher alcohol dependence and greater frequency of alcohol-related aggression. (Moeller & Dougherty, 2001).

Prolonged, excessive use of alcohol has been linked to many cognitive deficits which again may contribute to increased behavioral impulsivity, frustration, poor judgement, and ultimately aggression. In a retrospective study of Australian alcoholics (n = 641), almost 60% had frontal-lobe dysfunction as measured by neuropsychological tests, while 30% had short-term memory loss (Tuck & Jackson, 1991). It should be noted that in addition to alcoholism, many of the participants in this study had other neurological disorders, including cerebellar degeneration, peripheral neuropathy, and seizures. In a study of detoxified alcoholics, participants' verbal and nonverbal performance was impaired, and they had memory deficits and perceptual motor difficulties (Tivis, Beatty, Nixon, & Parsons, 1995). The extent of observed neuropsychological deficits seen in alcohol abusers is related to the severity of alcohol dependence (Keenan, O'Donnell, Sinanan, & O'Callaghan, 1997). Also, when alcohol is combined with other regular drug use, the extent of neuropsychological deficit increases. For example, polysubstance abusers show greater recall deficits on a verbal-learning task, although recognition rates are equivalent between alcohol and polydrug users (Bondi, Drake, & Grant, 1998).

While alcohol abuse is certainly linked to aggression, so too are other substances. Almost 14% of jailed violent offenders were under the influence of drugs (excluding alcohol) at the time of their offense (Department of Justice, 1996). Also, in a community sample, persons with a history of drug dependence showed more aggressive responses in a game using the Point Subtraction Aggression Paradigm than those without a substance abuse history (see Allen, Moeller, Rhoades, & Cherek, 1997). There seems to be an increased risk of aggression in substance abusers who also have a psychiatric disorder (Soyka, 2000). As with alcohol abusers, there may be a biological explanation for the aggressive behavior seen in drug users; that is, substance users who exhibit violent behavior may have compromised dopaminergic and serotonergic systems (Fishbein, 1998; see Chapter 9). The use of drugs and alcohol clearly has a profound effect on the incidence of aggressive behavior.

Other Psychiatric Conditions

The relationship between some psychiatric disorders, such as schizophrenia, and aggression is a controversial one. While the literature reviewed here suggests that schizophrenic patients are not necessarily more prone to violence, some researchers would disagree (e.g., Tardiff, 1998). There are, however, several psychiatric diagnoses that clearly involve aggression. In intermittent explosive disorder (IED), patients describe an irresistible impulse to be aggressive, and experience a sense of tension that is relieved by the explosive episode (McElroy, Soutullo, Beckman, Taylor & Keck, 1998). These patients also report an affective component to their behavior characterized by manic-like symptoms followed by depression. McElroy *et al.* also note that many of these patients have a childhood history of temper tantrums, impaired attention, hyperactivity, and other behavioral problems. Another disorder that very clearly involves externalizing behavior, conduct disorder is characterized by aggression towards people and animals, destruction of property, deceitfulness, and rule violation (DSM-IV; APA, 1994). Adolescents with conduct disorder have verbal deficits and show signs of executive dysfunction, including impaired attention, poor planning ability, and limited cognitive flexibility (Teichner & Golden, 2000). Antisocial Personality Disorder (ASPD) also involves aggression, with disregard for the welfare of others, lawlessness, and frequent physical altercations (DSM-IV; APA, 1994). Given that Conduct Disorder is a precursor of ASPD, it is presumed that some of the earlier cognitive deficits continue to be present in adulthood. Although it only occurs in extreme cases, executive dysfunction (Bearden, Hoffman, & Cannon, 2001) as well as aggression, anxiety, and psychosis are occasionally seen in manic patients (Swann, 1999). Aggression is also sometimes seen in patients with depression and dementia, although this behavior is not a primary symptom of either disorder (DSM-IV; APA, 1994). While most bipolar patients and most depressed patients are not aggressive, disturbed mood may contribute to greater impulsivity and irrational thinking, which may or may not be related to a cognitive deficit. Thus, while some psychiatric disorders, such as Antisocial Personality Disorder and Conduct Disorder, are clearly associated with aggression, others such as bipolar disorder, schizophrenia, and dementia appear to involve aggressive behavior only in rare circumstances.

ENVIRONMENTAL FACTORS

In addition to putative brain disease, researchers have also explored the role of numerous environmental factors as they relate to aggression. Environmental variables are relevant to aggression because they serve as a vulnerability and may also impact negatively on cognitive functioning. The scope of these environmental considerations is wide, and detailing them all is beyond the scope of this chapter; however, a few examples will be mentioned here. In a review article, Gartner and Whitaker-Azmitia (1996) note that lower socioeconomic status, abuse, neglect, and coercive family interactions are associated with aggressive behaviors in children. Similarly, in a study of youth violence, Dahlberg (1998) reviews several variables associated with aggression. These include early onset of aggressive behavior, social skills deficits, exposure to violence, poor parenting and family functioning, negative peer influences, access to guns, and living in neighborhoods characterized by poverty, transiency, family disruption, and social isolation. It does not appear than any one of these variables alone accounts for aggression. For example, in a study of adult males, individuals with either neurological disorders seen early in life, or environmental limitations (such as social or educational problems) had lower rates of criminal behavior, including aggressive acts, than did persons with both neurological deficits and unstable, conflict ridden family histories (Raine *et al.*, 1996). Although many environmental variables have been linked to aggressive behavior, these variables appear to work in conjunction with other limitations that also render an individual vulnerable, such as cognitive deficits and neurologic or psychiatric difficulties.

Parenting Style

Parenting style can also influence the emergence of aggressive behavior. In a prospective study of parenting style in a community sample, Newcomb and Loeb (1999) followed a group of women from childhood through parenthood and found that their use of a cold, rejecting, ineffective parenting style was associated with a prior childhood history of polydrug use, property crimes, and social nonconformity. Although Newcomb and Loeb did not assess the degree of aggressive behavior in the offspring of these women, in Raine *et al.*'s (1996) study, a history of maternal rejection and having a criminal parent were among the variables associated with violence in males. In a study of parents of aggressive toddlers, the authors found that maternal child-rearing practices, personality attributes, and drug use were related to aggression in 2-year olds (Brook, Zheng, Whiteman, & Brook, 2001). In a study of Chinese children, authoritarian parenting correlated positively with aggression and adjustment problems, while an authoritative style was associated positively with social and academic adjustment (Chen, Dong, & Zhou, 1997). Children who are spanked are more likely to behave aggressively towards peers, with more aggression seen in severely spanked children (Strassberg, Dodge, Pettit, & Bates, 1994). Finally, in a study of boys of low socioeconomic status, physical aggression was related to family adversity and poor parenting (Haapasalo & Tremblay, 1994). While it is impossible to separate parenting style from the many other variables in these studies (e.g., SES), the way in which parents interact with their children also has an impact on aggressive behavior.

Childhood Abuse

It has been noted that many aggressive individuals have been victims of some type of childhood abuse. A study of drug users suggests that many had victimized family members or sexual partners, and that many had been sexually and physically abused as children (Clarke, Stein, Sobota, Marisi, & Hanna, 1999). In a study of impoverished inner-city children attending a summer camp, maltreated children were more likely to be aggressive, and those who had experienced physical abuse, in particular, were more prone to reactive aggression (Shields & Cicchetti, 1998). Maltreated children also show more signs of impaired attention and emotion dysregulation (Shields & Cicchetti, 1998). The abuse need not be severe for problems with aggression and hostility to occur (Nicholas & Bieber, 1996). Finally, it seems that maltreated children are more likely to bully peers and also to be victimized by them, with emotion dysregulation being an important mediator of this cycle (Shields & Cicchetti, 2001). For some maltreated children, then, there is the possibility of future aggressive behavior as well as repeated victimization.

While a history of abuse may be associated with later aggression, it is unlikely that it is the only important factor. In his recent book, Pincus (2001) proposes a three-part explanation of violence in serial murderers, which includes a history of severe sexual and physical abuse along with brain damage and cognitive deficits. Serial murderers do not represent the larger population of people who were abused, many of whom have not had contact with the judicial system and who do not have problems with aggression. Although there are many likely reasons for this disparity, the number of white abused children exceeds the number of black abused children (Department of Health and Human Services, 1999), yet there are more African Americans imprisoned for violent crime (Department of Justice, 1996). Although childhood abuse may not be the only predictor of aggressive behavior, it is clearly a significant one, especially when one considers the psychological impact of such treatment.

Nutrition

Researchers have speculated about a possible link between dietary intake and aggressive behavior (see Kanarek, 1994). While it is difficult to parse out the impact of low socioeconomic status, hunger has been linked to aggression and increased anxiety in children, as well as other behavioral, emotional, and academic problems (Kleinman *et al.*, 1998). In a study of Kenyan children, activity level on the playground was related to food intake, but aggression seemed to be related to socioeconomic status only (Espinosa, Sigman, Bwibo, Neumann, & McDonald, 1992). Another nutritional variable that has been given considerable attention is sugar intake. Although many researchers have wondered about the effects of sugar on behavior, the overwhelming consensus seems to be that sugar is linked to aggressive behavior only in a small number of persons (Gans, 1991; Kanarek, 1994). In a study comparing adjudicated adolescent males with middle-school adolescents, a glucose tolerance test revealed a lower mean blood glucose nadir in the delinquent group, suggesting that glucose metabolism may be related to aggression in some individuals. Finally, although hypoglycemia is known to arouse angry feelings in some individuals (McCrimmon, Ewing, Frier, & Deary, 1999), this does not mean that these individuals will act on these feelings. Additional well-controlled studies are needed before any firm conclusions can be drawn, but at this point there does not appear to be a strong link between sugar and aggression. However, an association between hunger and aggression may exist.

Environmental Neurotoxins

Exposure to heavy metals, such as copper, zinc, and lead, is associated with aggression in animals and humans. Consistent with reports of children exposed to lead, such exposure in golden hamsters appears to be dose-dependent and is associated with lower birth weight and more aggressive behavior (Delville, 1999a, 1999b). In the Holtzman strain of rats, exposure to cadmium resulted in increased emotional reactivity and more intense frustration reactions (Davis, Arb, & Huss, 1995). In humans, elevated levels of copper and zinc were observed in a sample of young assaultive males (Walsh, Isaacson, Rehman, & Hall, 1997). Finally, in a well-designed prospective study, high lead levels were associated with anxious/depressed behavior, somatic complaints, impaired attention, aggression, and delinquency in 11 year old children (Needleman *et al.*, 1996). While exposure to metals is, again, one variable among many that is linked to aggression, Rice (1998) notes that even minimal effects in a few members of the population can have broader societal implications, including increased incidence of delinquency and antisocial behavior.

OTHER VARIABLES

In addition to the many variables mentioned thus far, the interplay between genetics and neurobiology is also important (see Chapters 3 and 4 of this volume). Twin studies suggest that genes play a significant role in violence and aggression (e.g., Alsobrook & Pauls, 2000; Seroczynski, Bergeman, & Coccaro, 1999). Also, a polymorphism found in the gene coding for tryptophan hydroxylase (TPH) was observed in a community sample (Manuck *et al.*, 1999). Men with the polymorphism in question scored higher on measures of aggression and reported more expression of anger than persons without this allele. With regard to neurobiology, many neurotransmitter systems have been studied, with particular emphasis on serotonin (see Miczek, Haney, Tidey, Vivian, & Weerts, 1994; Miczek, Mirsky, Carey, DeBold, & Raine, 1994; Chapter 9). Reduced activity in serotonin receptors is associated with a reduction in aggressive behavior in mice (Bell & Hobson, 1994), and a similar effect is thought to occur in humans. In a review article, Lesch and Merschdorf (2000) suggest that variability in serotonergic gene expression is at least partly responsible for aggressive behavior. Several structures of the limbic system, including the amygdala, have also been implicated in aggression (see also Chapter 7). For example, defensive cats—those that are especially reactive to environmental threats and anomalies—show larger transynaptic excitatory-evoked activity in the ventromedial hypothalamus and reduced evoked activity in the ventral hippocampus (Adamec, 1991). Psychosurgery studies also provide a clue to structures associated with aggressive behavior. Although some studies suggest that amygdalotomy is not an effective way to reduce aggressive behavior, two patients, one of whom had diffuse brain damage and epilepsy and the other of whom had limbic epilepsy, experienced a reduction in aggressive symptoms after amygdala-hippocampectomy (Sachdev, Smith, Matheson, Last, & Blumbergs, 1992). In a study of schizophrenic patients with a history of dangerous aggressive behavior, prefrontal bilateral lobotomy successfully eliminated or reduced offensive aggression in 87% of patients (n = 16; da Costa, 1997). Also, a reduction in aggressive behavior was observed in a schizophrenic patient who underwent two frontal lobotomies. (Mirsky & Rosvold, 1990). Finally, the role of hormones in aggression has also been explored. Although there is some disagreement, testosterone does not appear to contribute significantly to aggressive behavior (Albert, Walsh, & Jonik, 1993; Rubinow & Schmidt, 1996). In a study of gonadectomized monkeys, Mirsky (1955) concluded that gonadal hormones do not affect dominance patterns in either male or female macaques, nor that dominance in social situations appears to be a "hormonally conditioned behavior." Given that males commit more acts of violence than do females, exploring the effects of testosterone seems like a natural progression. However, there is little evidence to suggest that testosterone plays a central role in aggressive behavior.

A WORD ABOUT RESILIENCE

Given the number of variables thought to be related to aggression, the question arises as to why aggression and violence are not more prevalent. One possible explanation is *resilience* (see Sandler, 2001) which appears to be a common characteristic of many children, and serves to protect otherwise vulnerable persons from adversity (Masten, 2001). One way to enhance resilience is through support from caring adults (Wolkow & Ferguson, 2001). A case example of resilience, and one with which Americans are all familiar, is seen in the nation's response to the September 11 terrorist attacks. While these attacks involved unspeakable violence, people gathered together to comfort one another, and to aid in rescue and clean-up efforts. An article in the New York Times indicates that residents of

New York City are shaken, but possess a strong sense of resilience and a wary sense of hopefulness (Barry & Connelly, 2001). The subway crime rate in New York City has actually dropped since the attacks, suggesting that aggression does not always beget more aggression. A similar phenomenon is seen in the adaptation of the human brain in children, who sometimes recover remarkably after brain injury. Given the human capacity for resilience and adaptation, exploring these variables is also relevant to studies of aggression. Some children, like the group of Fetal Alcohol Syndrome adolescents identified in Streissguth's work, are able to thrive despite having multiple risk factors. Garmezy (1993) identified three factors that seem to protect children from adversity (see also Garmezy, 1985). These include the dispositional attributes of the child, such as intelligence and sociability, emotional support from a family member, and external support from other institutions such as school or work. With the aid of a caring adult and institutional support, a propensity towards aggressive behavior in high-risk individuals might be minimized or thwarted.

SUMMARY AND CONCLUSION

Many variables have been linked to aggressive behavior. These include environmental factors such as family instability (Raine *et al.*, 1996), low socioeconomic status (Gartner & Whitaker-Azmitia, 1996), inadequate nutrition (Kleinman *et al.*, 1998), physical or sexual abuse (Shields & Cicchetti, 1998), exposure to environmental neurotoxins, such as lead (Needleman *et al.*, 1996), cognitive deficits such as impulsivity, poor planning ability, working memory deficits, and low verbal intelligence (see Filley *et al.*, 2001; Séguin *et al.*, 1995), rare cases of psychiatric disturbance (e.g., schizophrenia; Spearing, 1999), prenatal exposure to alcohol (Olson *et al.*, 1997) neurobiological factors (e.g., Adamec, 1991) and genetic contributions (e.g., Alsobrook & Pauls, 2000). Cognitive deficits, especially impulsivity, poor planning ability, mental inflexibility, low verbal intelligence, and impaired attention, may limit an individual's ability to cope with other biological and environmental vulnerabilities. Such deficits may lead as well to increased feelings of frustration and anxiety, resulting in difficulty with regulation of emotion, and ultimately to aggression.

Cognitive compromise, especially in executive functioning, is observed in many aggressive individuals and serves as a vulnerability to and cause of aggressive behavior. The effects of these deficits are compounded by the presence of other vulnerabilities. As cognitive strength decreases, so too does the capacity to manage other stressors, leading to an increase in aggression and violence. In animal studies, stressors can impair attention and influence decision-making speed (Mendl, 1999). Similar effects are seen in humans, and might be more pronounced in individuals with limited cognitive skills. Managing cognitive deficits might be even more difficult when possible limbic-system damage is also a factor. Thus, brain-injured patients often experience a sense of intense frustration and irritability as a result of their injury and cognitive decline (see Stratton & Gregory, 1994). Even greater irritability might be expected in persons with cognitive limitations who also have limbic system compromise. Problems resulting from cognitive limitations might reduce one's ability to successfully and civilly manage a complex world. These include repeated academic failure and possible humiliation and frustration related to poor performance, coupled with neurologic or psychiatric disorders, abuse, and poverty. Those with brain disorders, cognitive limitations, and environmental hardships may find that the skills they do possess are not sufficient to overcome these obstacles. This may lead to aggression.

Although generally not accepted, and although it can have very grave consequences, aggression may be adaptive in some circumstances (Koolhaas *et al.*, 1999). This results from the increase in visibility afforded the aggressive person. Unfortunately, because parents, schools, and other institutions are themselves often operating with limited resources, they may not be able to provide the adequate support needed to address an aggressive individual's basic needs. When such intervention is available, some of the burden is lifted—thereby allowing the person to manage everyday demands more readily. When such intervention is not available, or when it is inadequate, the individual may resort to becoming increasingly aggressive. Although it is difficult to empathize with an aggressive child or adult, it is important to consider the cognitive, environmental, and biological context that might be driving the behavior. Children with limited cognitive abilities, who have other environmental stressors, such as family instability or a history of physical and sexual abuse, might be identified at an early age and given additional social and institutional support. As Garmezy's (1993) work on protective factors suggests, increased institutional and emotional support might increase a child's ability to successfully weather adversity, thereby reducing problem behaviors, including aggression and violence. Implementing this type of early support is in the long run less costly and more humane than constructing more prisons and increasing the size of the legal apparatus necessary to apprehend and punish violent offenders.

REFERENCES

American Psychiatric Association (1994). *DSM-IV: Diagnostic and statistical manual of mental disorders* (4th ed.). Washington, DC: Author.

Adamec, R.E. (1991). Individual differences in temporal lobe sensory processing of threat stimuli in the cat. *Physiology and Behavior*, **49**, 455-464.

Albert, D.J., Walsh, M.L., & Jonik, R.H. (1993). Aggression in humans: What is its biological foundation? *Neuroscience and Biobehavioral Reviews*, **17**, 405-425.

Allen, T.J., Moeller, G., Rhoades, H.M., & Cherek, D.R. (1997). Subjects with a history of drug dependence are more aggressive than subjects with no drug use history. *Drug and Alcohol Dependence*, **46**, 95-103.

Alsobrook, J.P., & Pauls, D.L. (2000). Genetics and violence. *Child and Adolescent Psychiatric Clinics of North America*, **9**, 765.

Babinski, L.M., Hartsough, C.S., & Lambert, N.M. (1999). Childhood conduct problems, hyperactivity-impulsivity, and inattention as predictors of adult criminal activity. *Journal of Child Psychology & Psychiatry*, **40**, 347-355.

Baird, J., Stevenson, J.C., & Williams, D.C. (2000). The evolution of ADHD: A disorder of communication. *The Quarterly Review of Biology*, **75**, 17-35.

Barry, D., & Connelly, M. (2001). A nation challenged survey: New Yorkers are shaken, but resolute, after attack. *The New York Times, Metropolitan Desk Section.* Avail. at www.nytimes.com.

Bassarath, L. (2001). Conduct Disorder: A biopsychosocial review. *Canadian Journal of Psychiatry*, **46**, 609-616.

Bearden, C.E., Hoffman, K.M., & Cannon, T.D. (2001). The neuropsychology and neuroanatomy of bipolar affective disorder: A critical review. *Bipolar Disorders*, **3**, 106-150.

Bell, R., & Hobson, H. (1994). 5-HT1A receptor influences in rodent social and agonistic behavior: A review and empirical study. *Neuroscience and Behavior*, **18**, 325-338.

Bergvall, A.H., Wessely, H., Forsman, A., & Hansen, S. (2001). A deficit in attentional set-shifting of violent offenders. *Psychological Medicine*, **31**, 1095-1105.

Bondi, M.W., Drake, A.I., & Grant, I. (1998). Verbal learning and memory in alcohol abusers and polysubstance abusers with concurrent alcohol abuse. *Journal of the International Neuropsychological Society*, **4**, 319-328.

Brook, J.S., Zheng, L., Whiteman, M., & Brook, D.W. (2001). Aggression in toddlers: Associations with parenting and marital relations. *Journal of Genetic Psychology*, **162**, 228-241.

Chen, X.Y., Dong, Q., & Zhou, H. (1997). Authoritative and authoritarian parenting practices and social and school performance in Chinese children. *International Journal of Behavioral Development*, **21**, 855-873.

Clarke, J., Stein, M.D., Sobota, M., Marisi, M., & Hanna, L. (1999). Victims as victimizers: Physical aggression by persons with a history of childhood abuse. *Archives of Internal Medicine*, **159**, 1920-1924.

Connor, P.D., Streissguth, A.P., Sampson, P.D., Bookstein, F.L., & Barr, H.M. (1999). Individual differences in auditory and visual attention among fetal alcohol-affected adults. *Alcoholism: Clinical and Experimental Research*, **23**, 1395-1402.

Cornwall A., & Bawden, H.N. (1992). Reading disabilities and aggression: A critical review. *Journal of Learning Disabilities*, **25**, 281-288.

da Costa, D.A. (1997). The role of psychosurgery in the treatment of selected cases of refractory schizophrenia: A reappraisal. *Schizophrenia Research*, **28**, 223-230.

Dahlberg, L.L. (1998). Youth violence in the United States: Major trends, risk factors, and prevention approaches. *American Journal of Preventative Medicine*, **14**, 259-272.

Davis, S.F., Arb, J.D., & Huss, M.T. (1995). Chronic cadmium exposure influences aggressive responding and partial reinforcement extinction performance. *The Psychological Record*, **45**, 565-575.

Deb, S., & Hunter, D. (1991). Psychopathology of people with mental handicap and epilepsy III: Personality disorder. *British Journal of Psychiatry*, **159**, 830-834.

Delville, Y. (1999a). Exposure to lead during development alters aggressive behavior in golden hamsters. *Neurotoxicology and Teratology*, **21**, 445-449.

Delville, Y. (1999b). Early lead exposure alters social communication and aggression. *Aggressive Behavior*, **25**, 7-8.

Department of Health and Human Services (2001). *The Numbers Count.* NIH Publication No 01-4584.

Department of Health and Human Services (1999). *10 Years of Reporting Child Maltreatment.* Children's Bureau.

Department of Justice (1996). *Profile of Jail Inmates.* Bureau of Justice Statistics.

Déry, M., Toupin, J., Pauzé, R., Mercier, H., & Fortin, L. (1999). Neuropsychological characteristics of adolescents with conduct disorder: Association with attention deficit hyperactivity disorder and aggression. *Journal of Abnormal Child Psychology*, **27**, 225-236.

Devinsky, O., Ronsaville, D., Cox, C., Witt, E., Fedio, P., & Theodore, W.H. (1994). Interictal aggression in epilepsy: The Buss-Durkee Hostility Inventory. *Epilepsia*, **35**, 585-590.

Elliott, D.S., & Tolan, P.H. (1999). Youth violence prevention, intervention, and social policy: An overview. In D.J. Flannery & C.R. Huff (Eds.), *Youth violence prevention, intervention, and social policy* (pp. 3-46). Washington, DC: American Psychiatric Press.

Espinosa, M.P., Sigman, M.D., Bwibo, N.O., Neumann, C.G., & McDonald, M.A. (1992). Playground behaviors of school-age-children in relation to nutrition, schooling, and family characteristics. *Developmental Psychology*, **28**, 1188-1195.

Farrington, D.P., & Loeber, R. (2000). Epidemiology of juvenile violence. *Child and Adolescent Psychiatric Clinics of North America*, **9**, 733.

Filley, C.M., Price, B.H., Nell, V., Antoinette, T., Morgan, A.S., Bresnahan, J.F., Pincus, J.H., Gelbort, M.M., Weissberg, M., & Kelly, J.P. (2001). Toward an understanding of violence: Neurobehavioral aspects of unwarranted physical aggression: Aspen Neurobehavioral Conference Consensus Statement. *Neuropsychiatry, Neuropsychology, & Behavioral Neurology*, **14**, 1-14.

Fishbein, D. (2000). Neuropsychological function, drug abuse, and violence: A conceptual framework. *Criminal Justice and Behavior*, **27**, 139-159.

Frith, C.D. (2000). The cognitive neuropsychology of schizophrenia. *International Journal of Psychology*, **35**, 272-273.

Gans, D.A. (1991). Sucrose and delinquent behavior: Coincidence or consequence. *Critical Review of Food Science*, **30**, 23-48.

Garmezy, N. (1993). Children in poverty: Resilience despite risk. *Psychiatry*, **56**, 127-136.

Garmezy, N. (1985). Stress-resistant children: The search for protective factors. In J.E. Stevenson (Ed.), *Recent Research in Developmental Psychopathology. Journal of Child Psychiatry Book Supplement*, **4**, 213-233.

Gartner, J., & Whitaker-Azmitia, P.M. (1996). Developmental factors influencing aggression. *Annals of the New York Academy of Sciences*, **794**, 113-120.

Goldberg, E., Bilder, R., Mattis, S., Antin, S., & Hughes, J. (1989). Reticulo-frontal disconnection syndrome. *Cortex*, **25**, 687-695.

Golden, C.J., Jackson, M.L., Peterson-Rohne, A., & Gontkovsky, S.T. (1996). Neuropsychological correlates of violence and aggression: A review of the clinical literature. *Aggression and Violent Behavior*, **1**, 3 -25.

Grassi, L., Peron, L., Marangoni, C., Zanchi, P., & Vanni, A. (2001). Characteristics of violent behaviour in acute psychiatric in-patients: a 5-year Italian study. *Acta Psychiatrica Scandinavica*, **104**, 273-279.

Gyimah-Brempong, K. (2001). Alcohol availability and crime: Evidence from census tract data. *Southern Economic Journal*, **68**, 2-21.

Haapasalo, J., & Tremblay, R.E. (1994). Physically aggressive boys from ages 6 to 12: Family background, parenting behavior, and prediction of delinquency. *Journal of Consulting and Clinical Psychology*, **62**, 1044-1052.

Harvey, R.J., Fletcher, J., & French, D.J. (2001). Social reasoning: A source of influence on aggression. *Clinical Psychology Review*, **21**, 447-469.

Hermann, B.P., Schwartz, M.S., Whitman, S., & Karnes, W.E. (1980). Aggression and epilepsy: Seizure-type comparisons and high risk variables. *Epilepsia*, **22**, 691-698.

Herzberg, J.L., & Fenwick, P.B.C. (1988). The aetiology of aggression in temporal-lobe epilepsy. *British Journal of Psychiatry*, **153**, 50-55.

Higley, J.D. (2001). Individual differences in alcohol-induced aggression: A nonhuman-primate model. *Alcohol Research & Health*, **25**, 12-19.

Juhász, C., Behen, M.E., Muzik, O., Chugani, D.C., & Chugani, H.T. (2001). Bilateral medial prefrontal and neocortical hypometabolism in children with epilepsy and aggression. *Epilepsia*, **42**, 991-1001.

Kanarek, R.B. (1994). Nutrition and violent behavior. In A.J. Reiss, K.A. Miczek, & J.A. Roth (Eds.), *Understanding and preventing violence Vol. 2* (pp. 515-539). Washington, DC: National Academy Press.

Keenan, E., O'Donnell, C., Sinanan, K., & O'Callaghan, E. (1997). Severity of alcohol dependence and its relationship to neurological soft signs, neuropsychological impairment, and family history. *Acta Psychiatrica Scandinavica*, **95**, 272-276.

Kleinman, R.E., Murphy, J.M., Little, M., Pagano, M., Wehler, C.A., Regal, K., & Jellinek, M.S. (1998). Hunger in children in the United States: Potential behavioral and emotional correlates. *Pediatrics*, **101**, E31-E36.

Koolhaas, J.M., Korte, S.M., DeBoer, S.F., Van Der Vegt, B.J., Van Reenen, C.G., Hopster, H., DeJong, I.C., Ruis, M.A.W., & Blokhuis, H.J. (1999). Coping styles in animals: Current status in behavior and stress-physiology. *Neuroscience and Biobehavioral Reviews*, **23**, 925-935.

Krakowski, M., Volavka, J., & Brizer, D. (1986). Psychopathology and violence: A review of the literature. *Comprehensive Psychiatry*, **27**, 131-148.

Kremen, W.S., Seidman, L.J., Faraone, S.V., & Tsuang, M.T. (2001). Intelligence quotient and neuropsychological profiles in patients with schizophrenia and in normal volunteers. *Biological Psychiatry*, **50**, 453-462.

Lesch, K.P., & Merschdorf, U. (2000). Impulsivity, aggression, and serotonin: A molecular psychobiological perspective. *Behavioral Science and the Law*, **18**, 581-604.

Lyman, D., Moffitt, T., & Stouthamer-Loeber, M. (1993). Explaining the relation between IQ and delinquency: Class, race, test motivation, school failure, or self-control? *Journal of Abnormal Psychology*, **102**, 187-196.

Manuck, S.B., Flory, J.D., Ferrell, R.E., Dent, K.M., Mann, J.J., & Muldoon, M.F. (1999). Aggression and anger-related traits associated with a polymorphism of the tryptophan hydroxylase gene. *Biological Psychiatry*, **45**, 603-614.

Masten, A.S.S. (2001). Ordinary magic: Resilience processes in development. *American Psychologist*, **56**, 227-238.

McCrimmon, R.J., Ewing, F.M.E., Frier, B.M., & Deary, I.J. (1999). Anger state during acute insulin-induced hypoglycemia. *Physiology and Behavior*, **67**, 35-39.

McElroy, S.L., Soutullo, C.A., Beckman, D.A., Taylor, P., & Keck, P.E. (1998). DSM-IV intermittent explosive disorder: A report of 27 cases. *Journal of Clinical Psychiatry*, **59**, 203-210.

Mendez, M.F., Doss, R.C., & Taylor, J.L. (1993). Interictal violence in epilepsy. *Journal of Nervous and Mental Disease*, **181**, 566-569.

Mendl, M. (1999). Performing under pressure: Stress and cognitive function. *Applied Animal Behaviour Science*, **65**, 221-244.

Miczek, K.A., Haney, M., Tidey, J., Vivian, J., & Weerts, E. (1994). Neurochemistry and pharmacotherapeutic management of aggression and violence. In A.J. Reiss, K.A. Miczek, & J.A. Roth (Eds.), *Understanding and preventing violence Vol. 2* (pp. 245-514). Washington, DC: National Academy Press.

Miczek, K.A., Mirsky, A.F., Carey, G., DeBold, J., & Raine, A. (1994). An overview of biological influences on violent behavior In A.J. Reiss, K.A. Miczek, & J.A. Roth (Eds.), *Understanding and preventing violence Vol. 2* (pp. 1-20). Washington, DC: National Academy Press.

Mirsky, A.F. (1955). The influence of sex hormones on social behavior in monkeys. *Journal of Comparative and Physiological Psychology*, **48**, 327-335.

Mirsky, A.F., & Rosvold, H.E. (1990). The case of Carolyn Wilson: A 38-year follow-up of a schizophrenic patient with two prefrontal lobotomies. In E. Goldberg (Ed.), *Contemporary neuropsychology and the legacy of Luria* (pp. 51-75). Hillsdale, New Jersey: Lawrence Erlbaum Assoc, Inc.

Mirsky, A.F., Yardley, S.J., Jones, B.P., Walsh, D., & Kendler, K.S. (1995). Analysis of the attention deficit in schizophrenia: A study of patients and their relatives in Ireland. *Journal of Psychiatric Research*, **29**, 23-42.

Moeller, F.G., & Dougherty, D.M. (2001). Antisocial personality disorder, alcohol, and aggression. *Alcohol Research & Health*, **25**, 5-11.

Moffitt, T.E. (1990). Juvenile delinquency and attention deficit disorder: Boys' development trajectories from age 3 to age 15. *Child Development*, **61**, 893-910.

Needleman, H.L., Riess, J.A., Tobin, M.J., Biesecker, G.E., & Greenhouse, J.B. (1996). Bone lead levels and delinquent behavior. *Journal of the American Medical Association*, **275**, 363-369.

Newcomb, M.D., & Loeb, T.B. (1999). Poor parenting as an adult problem behavior: General deviance, deviant attitudes, inadequate family support and bonding, or just bad parents? *Journal of Family Psychology*, **13**, 175-193.

Nicholas, K.B., & Bieber, S.L. (1996). Parental abusive versus supportive behaviors and their relation to hostility and aggression in young adults. *Child Abuse & Neglect*, **20**, 1195-1211.

Olson, H.C., Feldman, J.J., Streissguth, A.P., Sampson, P.D., & Bookstein, F.L. (1998). Neuropsychological deficits in adolescents with fetal alcohol syndrome: Clinical findings. *Alcoholism: Clinical and Experimental Research*, **22**, 1998-2012.

Olson, H.C., Streissguth, A.P., Sampson, P.D., Barr, H.M., Bookstein, F.L., & Thiede, K. (1997). Association of prenatal alcohol exposure with behavioral and learning problems in early adolescence. *Journal of the American Academy of Child and Adolescent Psychiatry*, **36**, 1187-1194.

Pascualvaca, D.M., Anthony, B.J., Arnold, L.E., Rebok, L.E., Ahearn, M.B., Kellam, S.G., & Mirsky, A.F. (1997). Attention performance in an epidemiological sample of urban children: The role of gender and verbal intelligence. *Child Neuropsychology*, **3**, 13-27.

Peled, A., Geva, A.B., Kremen, W.S., Blankfeld, H.M, Esfandiarfard, R., & Nordahl, T.E. (2001). Functional connectivity and working memory in schizophrenia: An EEG study. *International Journal of Neuroscience*, **106**, 47-61.

Pincus, J.H. (2001). *Base instincts*. New York: W.W. Norton & Co., Inc.

Pineda, D.A., Ardila, A., Rosselli, M., Puerta, I.C., Meija, S., & Toro, M.C. (2000). Neurobehavioral characteristics of adolescents with behavioral dysregulation disorder. *International Journal of Neuroscience*, **101**, 133-156.

Pliszka, S.R. (1998). Comorbidity of attention-deficit/hyperactivity disorder with psychiatric disorder: An overview. *Journal of Clinical Psychiatry*, **59**, 50-58.

Raine, A., Brennan, P., Mednick, B., & Mednick, S. (1996). High rates of violence, crime, academic problems, and behavioral problems in males with both early neuromotor deficits and unstable family environments. *Archives of General Psychiatry*, **53**, 544-549.

Rice, D.C. (1998). Issues in developmental neurotoxicology: Interpretation and implications of the data. *Canadian Journal of Public Health*, **89**, S31-S36.

Robertson, G. (1988). Arrest patterns among mentally disordered offenders. *British Journal of Psychiatry*, **153**, 313-316.

Rubinow, D.R., & Schmidt, P.J. (1996). Androgens, brain, and behavior. *American Journal of Psychiatry*, **153**, 974-984.

Sachdev, P., Smith, J.S., Matheson, J., Last, P., & Blumbergs, P. (1992). Amygdalo-hippocampectomy for pathological aggression. *Australian and New Zealand Journal of Psychiatry*, **26**, 671-676.

Sandler, I. (2001). Quality and ecology of adversity as common mechanisms of risk and resilience. *American Journal of Community Psychology*, **29**, 19-61.

Séguin, J.R., Boulerice, B., Harden, P.W., Tremblay, R.E., & Pihl, R.O. (1999). Executive functions and physical aggression after controlling for attention deficit hyperactivity disorder, general memory, and IQ. *Journal of Child Psychology and Psychiatry*, **40**, 1197-1208.

Séguin, J.R., Pihl, R.O., Harden, P.W., Tremblay, R.E., & Boulerice, B. (1995). Cognitive and neuropsychological characteristics of physically aggressive boys. *Journal of Abnormal Psychology*, **104**, 614-624.

Seroczynski, A.D., Bergeman, C.S., & Coccaro, E.F. (1999). Etiology of the impulsivity aggression relationship: Genes or environment? *Psychiatry Research*, **86**, 41-57.

Shields, A., & Cicchetti, D. (2001). Parental maltreatment and emotion dysregulation as risk factors for bullying and victimization in middle childhood. *Journal of Clinical Child Psychology*, **30**, 349-363.

Shields, A., & Cicchetti, D. (1998). Reactive aggression among maltreated children: The contributions of attention and emotion dysregulation. *Journal of Clinical Child Psychology*, **27**, 381-395.

Soyka, M. (2000). Substance misuse, psychiatric disorder, and violent and disturbed behaviour. *British Journal of Psychiatry*, **176**, 345-350.

Spearing, M.K. (1999). *Schizophrenia.* Department of Health and Human Services, National Institute of Mental Health, NIH 99-3517.

Spreen, O., Risser, A.T., & Edgell, D. (1995). *Developmental neuropsychology.* New York: Oxford University Press.

Stanford, M.S., Greve, K.W., & Gerstle, J.E. (1997). Neuropsychological correlates of self-reported impulsive aggression in a college sample. *Personality and Individual Differences*, **23**, 961-965.

Steinert, T., & Froscher, W. (1994). Aggression in epilepsy. (Trans. T. Crump) *Nervenheilkunde*, **13**, 199-205.

Steinert, T., Sippach, T., & Gebhardt, R.P. (2000). How common is violence in schizophrenia despite neuroleptic treatment? *Pharmacopsychiatry*, **33**, 98-102.

Strassberg, Z., Dodge, K.A., Pettit, G.S., & Bates, J.E. (1994). Spanking in the home and children's subsequent aggression toward kindergarten peers. *Development and Psychopathology*, **6**, 445-461.

Stratton, M.C., & Gregory, R.J. (1994). After traumatic brain injury: A discussion of consequences. *Brain Injury*, **8**, 631-645.

Streissguth, A.P., Barr, H., Kogan, J., & Bookstein, F. (1997). Primary and secondary disabilities in fetal alcohol syndrome. In A.P. Streissguth & J. Kanter (Eds.), *The challenge of fetal alcohol syndrome: Overcoming secondary disabilities*. Seattle: University of Washington Press.

Streissguth, A.P., Martin, D.C., Martin, J.C., & Barr, H.M. (1981). The Seattle longitudinal prospective study on alcohol and pregnancy. *Neurobehavioral Toxicology and Teratology*, **3**, 223-233.

Sun, H.F. (2001). The relationship between two aspects of executive function performance at age 8 and the development of aggressive behavior over the elementary and middle school years. Unpublished Ph.D. dissertation, Johns Hopkins University School of Hygeine and Public Health.

Swann, A.C. (1999). Treatment of aggression in patients with bipolar disorder. *Journal of Clinical Psychiatry*, **60**, 25-28.

Tardiff, K. (1998). Unusual diagnoses among violent patients. *The Psychiatric Clinics of North America*, **21**, 567-576.

Tebartz van Elst, L., Woermann, F.G., Limieux, L., Thompson, P.J., & Trimble, M.R. (2000). Affective aggression in patients with temporal lobe epilepsy: A quantitative MRI study of the amygdala. *Brain*, **123**, 234-243.

Teichner, G., & Golden, C.J. (2000). The relationship of neuropsychological impairment to conduct disorder in adolescence: A conceptual review. *Aggressive and Violent Behavior*, **5**, 509-528.

Tivis, R., Beatty, W.W., Nixon, S.J., & Parsons, O.A. (1995). Patterns of cognitive impairment among alcoholics: Are there subtypes? *Alcoholism: Clinical and Experimental Research*, **19**, 496-500.

Tuck, R.R., & Jackson, M. (1991). Social, neurological, and cognitive disorders in alcoholics. *Medical Journal of Australia*, **155**, 225-229.

Walsh, W.J., Isaacson, H.R., Rehman, F., & Hall, A. (1997). Elevated blood copper/zinc ratios in assaultive young males. *Physiology and Behavior*, **62**, 327-329.

Wolkow, K.E., & Ferguson, H.B. (2001). Community factors in the development of resiliency: Considerations and future directions. *Community Mental Health Journal*, **37**, 489-498.

Chapter 6

BRAIN INJURY AS A CONTRIBUTORY FACTOR IN OFFENDING

Edgar Miller, Ph.D.
Center for Applied Psychology, University of Leicester

INTRODUCTION

The aim of this chapter is to explore the possible relationship between brain injury or damage, on the one hand, and offending, on the other. In doing so it is important to set out in a little more detail the exact focus of concern.

The present concern is with brain injury or damage or injury associated with anatomic disturbance or pathology in the brain which has been sustained at birth or later. This can be distinguished from psychophysiological research dealing with such things as electrodermal activity, heart rate and autonomic conditioning (e.g., Raine & Venables, 1988; Raine, Venables & Williams, 1996; see Chapter 8 for a review of this literature), where the link is to altered physiological processes. Altered physiological processes like changes in electrodermal activity, heart rate, EEG and the like, need not necessarily reflect any underlying anatomical abnormality (although it is always possible that they may do so). In practice, the condition associated with brain damage that has attracted most attention has been closed head injury as caused by severe blows to the head that do not penetrate the skull. However, other forms of acquired brain damage have also been implicated.

A second background point relates to the standpoint from which this chapter is written. It is certainly not based on the premise that all who commit criminal acts are brain damaged in some way—this would be a highly improbable assumption. Rather, the thesis being examined is that amongst those who have suffered certain forms of brain damage, there may be an enhanced propensity to offend. It is readily accepted that the causes of crime are complex and multi-faceted and that brain injury may be a factor in only a minority of those who offend.

Finally, there are many forms of crime and it is highly unlikely that all can be explained by the same pattern of influences. The factors that lead to a fatal stabbing in a barroom brawl are unlikely to be exactly the same as those leading a businessman to falsify accounts or misrepresent the financial position to potential investors. If brain injury plays a role in precipitating criminal behaviour, as will be seen below, this is likely to be in relation to a limited range of offences and particularly those linked to violence.

There are three major lines of evidence that have a bearing on this issue and these will be explored in turn.

NEUROPSYCHOLOGICAL IMPAIRMENTS

It is well established that lesions of the brain may produce a very wide range of psychological impairments. Because these have been described in a wide range of neuropsychological publications (e.g., Walsh & Darby, 1999) only some key aspects of immediate relevance will be very briefly outlined here.

Closed head injuries, for example, are well established as producing disturbances in cognitive functioning, of which impairments in memory are probably the most prominent (e.g., Richardson, 2000). In addition, head injuries can produce a number of changes in personality and related functions. These consist of such things as impulsivity and lack of foresight, an impaired ability to plan ahead, disinhibition and lack of tact (Lishman, 1998). Disturbed sexual behaviour, irritability and even overt aggression can also arise (e.g., Wood, 1984). It would be not at all surprising if people with these characteristics were more prone than others to indulge in antisocial behaviour including certain forms of criminal behaviour.

The personality changes which occur in head injury are also very similar to those exhibited by people with more localized lesions of the frontal lobes (Lishman, 1998; Walsh & Darby, 1999). In addition, the temporal lobes of the brain have been associated with the control of violent emotions and sexually-related behaviour (Walsh & Darby, 1999). Again, it could be that those with more localized damage to the frontal or temporal lobes might show an excess of certain forms of offending behaviour. Damage to other parts of the brain, such as the parietal and occipital lobes, produce psychological changes that are not so likely to precipitate antisocial behaviour.

That it is reasonable to suppose that damage to the brain of certain kinds might help to precipitate offending behaviour is no guarantee that it actually does do so. To establish the link properly requires much better evidence than has been provided so far. Much of the rest of this chapter will be directed towards analyzing the further evidence that is currently available. In passing, it can also be noted that it is also possible that the personality changes associated with head injury, which might tend to push the head injury victim towards committing acts of antisocial behaviour, may be counteracted by other factors which act in the reverse direction and so neutralize the effects of the personality changes. This may well be the case with very severe head injuries (associated with periods of unconsciousness of the order of a week or more). Whilst victims of such very severe injuries will typically display the personality changes associated with head injury as outlined above, they also often have a wider range of handicaps including physical impairments (e.g., hemiplegia). These may reduce the opportunity to offend since many forms of offending require a certain level of mobility.

STUDIES OF OFFENDER POPULATIONS

A large number of investigations have been carried out which attempt to ascertain rates of past head injury amongst offender populations, and especially groups of violent offenders. The populations surveyed have consisted of such groups as convicted murderers, prisoners on *death row*, and the inmates of special psychiatric facilities for psychiatrically disturbed offenders. The results of these studies are summarized in Table 1. It can be seen that the proportion with head injuries varies between 10% in Blake, Pincus and Buckner's (1995) report based on 31 murderers and 67% in the study by Lewis, Pincus, Feldman, Jackson and Bard (1986) of convicts on *death row*. Even the lowest of these figures is very much in excess of what epidemiological research would suggest is the prevalence of past head injury in the general population, which is of the order of 1-2% at most (Jennett & MacMillan, 1981).

Table 1. Prevalence of head injury in offender groups.

Study	N	Nature of group	Rate
Bach-y-Rita & Veno (1974)	62	Habitually violent offenders	61%
Blake *et al.* (1995)	31	Murderers	10%
DelBello *et al.* (1999)	25	Sexual Offenders	36%
Frierson *et al.* (1998)	54	Murderers	24%
Gibben *et al.* (1959)	72	Severely psychopathic criminals	40%
Lewis *et al.* (1986)	15	Convicts on "death row"	67%
Lewis *et al.* (1988)	14	Convicts on "death row" sentenced as juveniles	58%
Lumsden *et al.* (1998)	97	Consecutive admissions to UK special hospital	42%
Martell (1992)	50	Inmates in maximum security hospital for offenders	22%

At first sight, the evidence summarized in Table 1 appears to suggest a very clear link between head injury and offending since the recorded prevalence of head injury is very high. Unfortunately, there are a number of limitations surrounding investigations of this type which make it necessary to interpret the data with at least some degree of caution (Miller, 1999a). The most important of these will be described below.

Firstly, the determination of head injury in offender samples is not easy. Often access to records of past hospital admissions and the like can be difficult or even impossible to obtain, and this is especially so for those in penal establishments. Investigators then have to rely unduly on offenders' own self-reports. Some may well have reasons to exaggerate or even invent past injuries since this might be viewed as being potentially in the informant's own interest to do so (e.g., by offering a basis to argue for a reduced sentence on appeal). Linked to this, it is also then difficult to have very precise criteria as to what constitutes a head injury of significance, as opposed to a trivial bang on the head, due the problem of confirming details like the length of unconsciousness or length of post-traumatic amnesia. It is this latter type of information which will indicate whether the injury was serious enough to be likely to have caused underlying brain damage, and with likely long-term psychological consequences. All this is likely to mean that different studies have used data with different levels of reliability and different criteria as to what constitutes a significant head injury. This may explain some of the rather wide variation in the findings given in Table 1.

A second limitation is that it is difficult to know what the real comparison rate of head injury ought to be. Although it is possible to cite prevalence data relating to the whole population, it is well established that victims of head injury tend to be over-represented in certain sections of the population. Those most likely to suffer significant head injuries are young adult males from socially deprived backgrounds. This is very much the same group who are most prone to offend (e.g., Jennett & MacMillan, 1981; Richardson, 2000).

The prevalence rate for head injury against which it would be appropriate to compare rates in offender sample is therefore likely to be rather greater than the 1-2% indicated above. It is very difficult, on the basis of existing epidemiological data for head injury, to determine exactly how much greater this comparison figure ought to be. Nevertheless, making all due allowance for this point, the reported rates of between 10% and 67% in offender groups still seem rather high.

A third difficulty arises out of the fact that the offender groups studied have all been rather unusual groups (those on *death row*, inmates in special secure psychiatric units, etc.). They are certainly typical samples of offenders, even violent offenders, who would be convicted by the courts. One possibility is that having suffered a head injury has little impact on whether an individual may offend. The head injury may exert an effect, in that those who exhibit the associated personality and other changes and who happen to offend, may then be more likely to be placed in secure psychiatric facilities, or to find their way on to *death row* if they have committed murder.

The one investigation with information of relevance on this latter point is Frierson, Schwartz-Watts, Morgan and Malone (1998), who studied murderers and had a subgroup who had been given death sentences, and another awarded other sentences. There was no appreciable difference in the prevalence of head injury in the two groups. This one study, therefore, offers no support to the notion that a past head injury might solely act to affect disposal or sentencing once someone had been convicted rather than being directly linked to the offence in the first place.

A final problem with data like that in Table 1 is that whilst these do suggest an association between head injury and offending, they do not offer any good evidence as to the direction of causation. Whilst it is plausible to assume that head injury might act to precipitate certain forms of offending, it is also the case that indulging in violent activities might make those who do so prone to suffer head injuries.

In consequence head injury could be the cause or consequence of violent offence-related behaviour. It is also possible that causation acts in both directions. In some instances, head injury may help to precipitate offending whilst, in others, it is the consequence of indulging in violent activities. An indication that head injury may assist in producing offending is provided by examples of individuals whose offending career started later in life, and only after they had sustained a head injury (Sarapata, Herrmann, Johnson & Aycock, 1996). A later section will present better evidence indicating that head injury can play a causal role, although this does not rule out the possibility that it may be a consequence for some violent offenders.

On balance, the evidence relating to head injury considered so far, does point to a link between head injury and offending, especially for offences involving violence. The limitations in this evidence do need to be kept in mind and limit the confidence with which this conclusion can be drawn. Assuming that the relationship does exist, it raises other issues such as the direction of causation, which remain to be properly explored.

The Effects of Focal Lesions

Although closed head injuries may produce areas of especially marked contusion (the frontal lobes being one common site), damage to the brain is always diffuse and more-or-less all parts of the brain are affected to some degree (Teasdale & Mendelow, 1981). Consideration will turn to the effects of more anatomically restricted and focal lesions.

As has been indicated, the personality changes following head injury are very similar to those encountered in more localized frontal lesions. It would therefore be expected that similar evidence might exist for links between offending behaviour and frontal-lobe damage.

There is evidence suggesting that this is so, and Berman and Siegel (1976) were early advocates of the idea that the frontal lobes might be implicated in offending.

Gorenstein (1982) administered a set of neuropsychological tests, such as the Wisconsin Card Sorting and Stroop Colour-Word Interference tests, known to be sensitive to frontal impairment, to both psychopathic and non-psychopathic offenders. The psychopaths were noted to perform less well on many of the measures used. Hare (1984) carried out a more sophisticated and better controlled study of a similar kind, which used more rigorous criteria to classify participants, and matched groups more carefully for such things as age and intelligence. He failed to confirm Gorenstein's (1982) basic finding. In yet a further investigation, Devonshire, Howard, and Sellars (1988) found that secondary psychopaths (according to the definition offered by Blackburn, 1974), but not primary psychopaths, performed poorly on one of the most commonly used tests of frontal lobe impairment, the Wisconsin Card Sorting Test. This study used Blackburn's (1974) criteria to distinguish between primary and secondary psychopaths who show similar characteristics, apart from the former being extraverted and the latter withdrawn.

Other investigators have examined groups of offenders differentiated as to the nature of the offence. Both Bryant, Scott, Golden and Tori (1984) and Moffitt and Henry (1989) found that it was those committing offences involving aggression who were most prone to exhibit frontal type impairments on neuropsychological tests. Bryant *et al.* (1984) also found patterns of neuropsychological test performance consistent with temporal lobe lesions.

Investigations of this type suffer from two major limitations. Firstly, brain pathology has to be inferred from neuropsychological test performance and there is no independent evidence of underlying brain damage. It is always possible that performance on any neuropsychological test can decline for reasons other than underlying brain pathology. Secondly, even if poor performance on tests sensitive to frontal damage is really associated with underlying pathology, there is no guarantee that the damage is restricted to the frontal lobes of the brain. Many of those with closed head injuries and resultant diffuse brain pathology, would perform poorly on frontal tests. Moving on from these two points, it would be desirable if future research of a similar kind could obtain better indications of the existence and extent of any brain pathology (e.g., by using various brain-imaging techniques, such as magnetic resonance imaging).

Other evidence that frontal lobe malfunctioning might be specifically implicated comes from experiments using a form of brain scanning (PET or positron emission tomography). Raine, Buchsbaum *et al.* (1994) found reduced glucose metabolism in murderers pleading not guilty by reason of insanity as compared to a normal control group. Murderers did not differ from controls in terms of glucose metabolism in the temporal lobes, which indicates that the effect might be specific to the frontal lobes. Raine, Buchsbaum and La Casse (1997) were able to double the original sample of murderers (from 22 to 41), and the same findings were still present. This work is, however, limited to those investigated after pleading not guilty by reason of insanity, and for many of whom a question of possible brain damage had already been raised.

Considering the evidence overall, there are some grounds for linking frontal brain lesions to offending, and this is especially so for violent offending. There is also a hint as to the relevance of temporal lobe lesions. Again really firm conclusions are not possible due to methodological difficulties. A more extensive use of the various brain-imaging methods in studies of offenders might help to clarify some of the outstanding questions.

Lateralization Effects

The suggestion that anomalous patterns of lateralization within the brain might produce problems has been put forward in relation to a wide range of disorders, including developmental dyslexia and schizophrenia. Findings in these other contexts have typically been contradictory (e.g., Eglinton & Annett, 1994; Miller, 1983).

The notion that criminals suffer an imbalance in the cerebral hemispheres has a long history, going back to Lombroso in the late nineteenth century (Nachson, 1986; see Chapter 1). Unfortunately, the findings in relation to offender samples have often been somewhat inconclusive or even inconsistent, matching the outcome of research into lateralization and psychopathology, and dyslexia. One approach has been to look at patterns of handedness, since these bear some relations to lateralization of function within the brain. For example, Taylor, Brown and Gunn (1983) found that *neurotic prisoners* and violent offenders had a significantly greater proportion with strongly right-handed tendencies than controls. In contrast, Fitzhugh (1973) claimed an excess of left handers amongst delinquents, whilst Spellacy (1977) found no differences in handedness between violent and non-violent adolescents.

Other investigations have relied on the notion that those with left-hemisphere lesions tend to do less well on the verbal than on the performance subtests of the Wechsler Adult Intelligence Scale (in its original and revised versions), whereas right-hemisphere damage produces the reverse pattern of subtest performance. The general pattern of findings has been that samples of delinquents or older offenders do generally perform less well on the verbal than the performance subtests of the Wechsler intelligence tests, although there are a few contrary findings. The relevant studies are reviewed in detail by Moffitt (1990) and Nachson & Denno (1987).

A more experimental methodology was used by Hare and Connolly (1987) who described a set of experiments relating to lateralization. One of these used stimuli (words) presented tachistoscopically to left or right visual fields. The dependent variable was the time to recognize the words, and a normal control group showed superior recognition for words presented to the right visual field (which would be initially processed by the left hemisphere). A group of psychopaths showed the reverse effect, which suggests that the pattern of language dominance in the brain might be different for this group. Another one of Hare and Connolly's (1987) experiments, using the technique of dichotic listening, showed trends which were not as marked but which were still consistent with the tachistoscopically-based experiment.

If differences in lateralization are presumed to reflect damage to one hemisphere, which is one presumption on which much of this work is based, then work on lateralization in offenders suffers from a number of substantial limitations. These not only relate to the fact that the findings are far from wholly consistent, which is in itself a qualification that needs to be kept in mind.

In the first place, determining lateralization in the brain from differences between the verbal and performance subtests of the Wechsler intelligence scales is a method that is subject to question (Lezak, 1983; Miller, 1983). This is on the grounds that the underlying trends are weak, and discrepancies between the verbal and performance subtests can arise for reasons unrelated to any underlying differences or pathology in the brain. One example being, that the verbal subtests are much more dependent on the predominantly verbal skills acquired through education and delinquents, and that offenders are less likely to have received adequate schooling (Farrington, 1997).

From the point of view of the major concern of this chapter, all the work on lateralization, including that based on intelligence test performance, also shares a common limitation with that based on putative focal lesions. Altered patterns of lateralization may possibly reflect underlying brain damage or pathology, but need not necessarily do so. There are people among the ordinary population who do not follow the usual pattern of lateralization of functions within the brain, that is, by being right-dominant or of mixed dominance for language (Annett, 1985), and it is possible that there are consequences of left-handedness (with right dominance for language) which might affect the liability to commit criminal acts, or that some factor which predisposes to left handedness might also increase the risk of offending.

CLINICALLY-BASED STUDIES

In the previous section, investigations were carried out based on groups identified in terms of their criminal status, and then looking for indications of head injury or other possible indices of brain injury or pathology. An alternative approach is to start with groups defined in terms of their clinical characteristics, and then to determine the propensity of such groups to offend. This approach is also more directly linked to the aims of this chapter, which is to consider whether certain forms of brain injury may predispose to offending rather than attempt to demonstrate that offenders tend to be brain damaged.

There has been extensive research into the outcome after closed head injury in terms of such things as persistence of memory impairments, occupational and social adjustment, and the like. Within quite this extensive body of work, Miller (1999b) managed to identify a small number of follow-up studies which have reported upon offending or offence-related incidents (e.g., being arrested), both before and after head injury. These reports are summarized in Table 2.

As can be seen from Table 2, a number of different indices were used, from Brooks, Campsie, Symmington, Beattie and McKinlay's (1986) vaguely defined "in trouble with the law," through arrests, convictions, and up to incarceration. The mean ages of the samples at the time of the head injury was mid- to late-twenties and the follow-up periods ranged from 2 to 5 years. Without exception, the data actually reported refer to the number of incidents prior to the time of the head injury, and the number during the follow-up period. This does not offer a suitable basis to compare offence-related behaviour both before and after injury. For this reason, the data in Table 2 are based on calculating the mean percentage involved per year at risk, assuming that all pre-injury offending would have taken place after reaching the age of 18 years. Given the general pattern of offending across the age range (Farrington, 1997), this compresses the pre-injury offence-related behaviour into a rather narrower age range than is likely to be the case and so will overestimate the pre-injury figures.

Table 2. Offence-related activity before and after head injury.

Study	Offence related criterion	% per year Pre	Post
Brooks *et al.* (1986)	"In trouble with the law"	0.8	6.2
Kreutzer *et al.* (1991)	Arrests	2.5	1.6
Hall *et al.* (1994)	Arrests	2.0	4.0
	Incarcerations	1.0	5.0
Kreutzer *et al.* (1995)	Arrests	1.3	3.6
	Incarcerations	1.0	2.5

It can be seen that the rate of offence-related events increases in five out of six of the sets of data. The only exception is for the arrest rates reported by Kreutzer, Wehman, Harris, Burns and Young (1991). This one finding that appears to go against the trend may be, at least partly, because the pre-injury data exaggerate the pre-injury rate, as explained above, by assuming that no arrests will have taken place before the age of 18 years. Evidence like that provided by Farrington (1997) would indicate that this is unlikely to have been the case, and that some arrests at an earlier age may well have occurred. A second relevant factor is that for most offenders, the time of peak offending is prior to 20 years of age, and thereafter the rate of offending tends to decline. What this means is that if the head injuries had no effect at all on offending, the post-injury rates should be clearly lower than the pre-injury rates.

Putting these two points together makes the Kreutzer *et al.* (1991) data much less compelling as an exception to the general trend than might appear at first sight. In fact, with three studies reporting the number of arrests, it is possible to combine their data to calculate a weighted average of pre- and post-injury arrest rates. These emerge as being 1.6% per year for pre-injury and 3.3% per year after the head injury. A doubling of the arrest rate after a head injury, under circumstances when it would be otherwise expected to decline, does quite strongly suggest that head injury acts to enhance the likelihood of offending. Conversely, it is not likely to be the result of indulging in violent behaviour in most instances, although it is still possible that a minority of offenders may suffer brain injury as a result of their violent activities.

Kreutzer *et al.* (1991) and Kreutzer, Marwitz and Witol (1995) also draw attention to a potentially contaminating factor. Their cases of head injury had high rates of substance abuse (both alcohol and illicit drugs). However, it was also the case that the rate of substance abuse dropped quite markedly after head injury, which supports the notion that it was the head injury that helped raise the rate of post-injury, offence-related incidents and not the substance abuse.

Two other reports merit specific comment. Ommaya *et al.* (1996) examined discharge from military service of a very large sample of service personnel. The rate of discharge for behavioural reasons (which could include some offending) was 1.8 times more likely in those who had suffered mild head injury as opposed to those who had not suffered a head injury. After mild head injury, the chances of being discharged following a conviction against military law was raised by a factor of 2.7. In contrast, those who had suffered severe head injuries were very much more likely to be discharged for medical reasons, presumably because they would have been much more obviously handicapped and possibly multiply handicapped. Their opportunity to remain in the service after injury and to offend against the military system was probably much reduced.

Although this is evidence which again supports the notion that head injury (at least within a certain range of severity) does increase the risk of offending, it would be unwise to assume that the size of the effect would match civilian samples. Military law tends to cover a range of things (e.g., refusal to obey legitimate orders) that would not usually reach the criminal justice system in civilian life. In fact, the whole nature of the military system might make it more sensitive to the impact of head injury.

Raine, Brennan and Mednick (1994) reported a long-term, follow-up study of a large sample of infants who had been born in maternity hospitals in Copenhagen, up to the age of 18 years. The outcome variable was having been recorded as being involved in violent crime by the age of 18 years, and this was examined in relation to an index of maternal rejection at one year old, and the occurrence of perinatal complications at birth. Perinatal complications are known to be linked to brain damage (a summary of relevant evidence is provided by Raine, 1993). A little less than 4% of those with neither birth complications nor maternal rejection had exhibited any form of violent offending, and this rate was not exceeded by those who had either of these characteristics alone. In contrast, the small proportion of the sample who had suffered both birth complications and maternal rejection had a rate of violent offending, at 18 years, which approached 9% (i.e., their rate had more than doubled).

This latter report is of interest for two reasons. Brain damage associated with perinatal complications must antedate offending, and cannot be a consequence of offending. Secondly, and also of some significance, it draws attention to the fact that the effects of brain injury probably do not act in isolation. They are likely to interact with other influences in leading to offending, particularly psychosocial factors (see Chapter 5). Other than this particular report, the interaction of brain injury with other types of factors in leading to offending remains to be explored.

In summary, the evidence considered in this section both confirms the link between brain injury and offending and points to the direction of causality. It appears that brain injury of certain types can predispose to offence-related behaviour. This does not exclude the possibility that indulgence in violent behaviour can itself lead to brain injury. The important point in the present context is that the existence of a causal link from brain injury to offending is quite strongly indicated.

CONCLUDING COMMENTS

The material examined in this chapter points towards three conclusions. Firstly, that certain forms of brain injury (e.g., closed head injuries and frontal lobe lesions) produce psychological changes which are likely to make the individuals concerned more likely to offend than would otherwise be the case. Secondly, selected offender groups and especially those who have committed offences involving violence, are likely to have a raised prevalence of head injury and possibly also frontal-lobe damage. There is just a hint that temporal-lobe lesions might also be associated with offending. Although this goes outside the scope of this chapter, there is also psychophysiological evidence that is consistent with the possibility of a temporal-lobe disturbance in some offenders (Miller, 1999a).

These conclusions appear reasonable, are consistent with the evidence and well enough established to guide further work. Despite this, their acceptance needs to be tempered by the fact that there are a number of methodological difficulties surrounding the work reviewed. These have been described at appropriate points and will not be repeated here. These difficulties mean that some further work to verify the conclusions, using more rigorous designs less open to alternative explanations, would be welcome

Assuming that the conclusions set out above are reasonably valid and not seriously contradicted by future research, some key questions still remain. Certain forms of brain damage may help to precipitate offending, but there is a large conceptual gap between anatomical damage to the brain, on the one hand, and criminal behaviour, on the other. The question arises as to what mediates this link. The most likely and obvious answer is that it is the psychological consequences of brain injury, especially changes in such things as personality and related functions produced by both head injury and frontal lobe lesions. Personality changes, such as disinhibition and irritability could offer a linking mechanism, as could some of the more cognitive impairments. The latter include the inability to plan ahead and other executive functions (e.g., the inability to modify established patterns of behaviour in the light of new circumstances) that have been a focus of much recent research on the neuropsychology of the frontal lobes (e.g., Roberts, Robbins & Weiskrantz, 1998). The question as to what might be the key psychological factors has not been addressed by research that has been carried out so far, and it is the exploration of possible links of this kind that needs to provide the focus for much future research in this area.

Within such investigations, it would also be desirable to explore how the relevant psychological changes produced by brain injury interact with other variables such as social factors. As indicated earlier, the work of Raine, Brennan and Mednick (1994) indicates that the effects of brain injury may interact with psychosocial factors. Not only should this work help in designing suitable interventions for those who suffer brain injury in order to help prevent offending, but it should also assist in identifying exactly who amongst those with head injuries or frontal lesions might be most prone to offend. This is because not all those with head injuries or frontal lesions show the different neuropsychological changes to the same degree. Some more specific neuropsychological profile of the features likely to be associated with offending might be developed.

There are also wider issues that might arise from neuropsychological research concerned with the link between brain injury and offending. One is the potential role that the psychological changes associated with brain injury might play in relation to mens rea and criminal responsibility, particularly the question of diminished responsibility. As Buchanan (2000) has demonstrated, the underlying issues here are extremely complex. These go beyond the scope of this chapter, but one example is the difficult debate within jurisprudence as to the circumstances under which certain things might be allowed to offer some sort of mitigation or excuse for the committing of criminal acts.

The final point to be made is that the neuropsychological aspects of offending or criminal behaviour remain poorly explored. This chapter demonstrates that neuropsychological issues may play a small, but not insignificant, part in the understanding and management of some forms of crime, and therefore need to be kept in mind by those with wider criminological concerns.

REFERENCES

Annett, M. (1985). *Left, right, hand and brain.* London: Lawrence Erlbaum

Bach-y-Rita, G., & Veno, A.R. (1974). Habitual violence; a profile of 62 men. *American Journal of Psychiatry*, **131**, 1015-1017.

Berman, A., & Siegel, A. (1976). A neuropsychological approach to the etiology, prevention and treatment of juvenile delinquency. In A. Davids (Ed.), *Child personality and psychopathology: Current topics.* New York: Wiley.

Blackburn, R. (1974). *Personality and the classification of psychopathic disorders.* Special Hospitals Research Report, No. 10. London: Specials Hospitals Research Unit.

Blake, P.Y., Pincus, J.H., & Buckner, C. (1995). Neurologic abnormalities in murderers. *Neurology*, **45**, 1641-1647.

Brooks, D.N., Campsie, L., Symmington, C., Beattie, A., & McKinlay, W. (1986). The five year outcome of severe blunt head injury. *Journal of Neurology, Neurosurgery and Psychiatry*, **49**, 764-770.

Bryant, E.T., Scott, M.L., Golden, C.J., & Tori, C.D. (1984). Neuropsychological deficits, learning disability and violent behavior. *Journal of Consulting and Clinical Psychology*, **52**, 323-324.

Buchanan, A. (2000). *Psychiatric aspects of justification, excuse and mitigation.* London: Jessica Kingsley Publishers.

DelBello, M.P., Soutullo, C.A., Zimmerman,M.E., Sax, K.W., Williams, J.R., McElroy, S.L., & Strakowski, S.M. (1999). Traumatic brain injury in individuals convicted of sexual offences with and without bipolar disorder. *Psychiatry Research*, **89**, 281-286.

Devonshire, P.A., Howard, R.C., & Sellars, C. (1988). Frontal lobe functions and personality in mentally abnormal offenders. *Personality and Individual Differences*, **9**, 339-344.

Eglinton, E., & Annett, M. (1994). Handedness and dyslexia: a meta-analysis. *Perceptual and Motor Skills*, **79**, 1611-1616.

Farrington, D. (1997). Human development and criminal careers. In M. McGuire, R. Morgan, & R. Reiner (Eds.), *The Oxford handbook of criminology* (2nd ed.). Oxford: Oxford University Press.

Fitzhugh, K.B. (1973). Some neuropsychological features of delinquent subjects. *Perceptual and Motor Skills*, **36**, 494.

Frierson, R.L., Schwartz-Watts, D.M., Morgan, D.W., & Malone, T. (1998). Capital vs non-capital murder. *Journal of the American Academy of Psychiatry and Law*, **26**, 403-410.

Gibben, T.C.N., Pond, D.A., & Stafford-Clark, D. (1959). A follow-up study of criminal psychopaths. *Journal of Mental Science*, **105**, 108-115.

Gorenstein, E.E. (1982). Frontal lobe functions in psychopaths. *Journal of Abnormal Psychology*, **91**, 368-379.

Hall, K.M., Kazmark, P., Stevens, M., Englander, J., O'Hare, P., & Wright, J. (1994). Family stressors in traumatic brain injury: a two-year follow-up. *Archives of Physical Medicine and Rehabilitation*, **75**, 876-884.

Hare, R.D. (1984). Performance of psychopaths on cognitive tasks related to frontal lobe function. *Journal of Abnormal Psychology*, **93**, 133-140.

Hare, R.D., & Connolly, J.F. (1987). Perceptual asymmetries and information processing in psychopaths. In S.A. Mednick, T.E. Moffitt & S.A. Stack (Eds.), *The causes of crime: New biological approaches* (pp. 211-218). Cambridge: Cambridge University Press.

Jennett, B., & MacMillan, R. (1981). Epidemiology of closed head injury. *British Medical Journal*, **282**, 101-104.

Kreutzer, J.S., Marwitz, J.J., & Witol, A.D. (1995). Interrelationships between crime, substance abuse and aggressive behaviours among persons with traumatic brain injury. *Brain Injury*, **9**, 757-768.

Kreutzer, J.S., Wehman, P.H., Harris, J.A., Burns, C.T., & Young, H.F. (1991). Substance abuse and crime patterns among persons with traumatic brain injury referred for supported employment. *Brain Injury*, **5**, 177-187.

Lewis, D.O., Pincus, J.H., Bard, B., Richardson, E., Pricheps, L.S., Feldman, M., & Yeager, C. (1998). Neuropsychiatric, neuroeducational and family characteristics of 14 juveniles condemned to death in the United States. *American Journal of Psychiatry*, **145**, 584-589.

Lewis, D.O., Pincus, J.H., Feldman, M., Jackson, L., & Bard, B. (1986). Psychiatric, neurological and psychoeducational characteristics of 15 death row inmates in the United States. *American Journal of Psychiatry*, **143**, 838-845.

Lezak, M.D. (1983). *Neuropsychological assessment* (2nd ed.). New York: Oxford University Press.

Lishman, W.A. (1998). *Organic psychiatry* (3rd ed.). Oxford: Blackwell Scientific.

Martell, D.A. (1992). Estimating the prevalence of brain dysfunction in maximum security forensic psychiatric patients. *Journal of Forensic Sciences*, **37**, 878-893.

Miller, E. (1983). Intellectual function and its disorders. In M. Shepherd & O.L. Zangwill (Eds.), *General psychopathology: Handbook of psychiatry*, Vol. 1. Cambridge: Cambridge University Press.

Miller, E. (1999a). The neuropsychology of offending. *Psychology, Crime & Law*, **5**, 297-318.

Miller, E. (1999b). Head injury and offending. *Journal of Forensic Psychiatry*, **10**, 157-166.

Moffitt, T.E. (1990). The neuropsychology of juvenile delinquency: a critical review. In N. Tonry & N. Morris (Eds.), *Crime and justice: A review of research* (Vol. 12). Chicago: Chicago University Press.

Moffitt, T.E., & Henry, B. (1989). Neuropsychological assessment of executive functions in self-reported delinquents. *Development and Psychopathology*, **1**, 105-118.

Nachson, I. (1986). Hemisphere function in violent offenders. In T.E. Moffitt & S.A. Mednick (Eds.), *Biological contributions to crime causation* (pp. 55-67). Dordrecht: Martinus Nijhoff.

Nachson, I., & Denno, D. (1987). Violent behavior and cerebral hemisphere function. In S.A. Mednick, T.E. Moffitt, & S.A. Stack (Eds.), *The causes of crime: New biological approaches* (pp. 185-217). Cambridge: Cambridge University Press.

Ommaya, A.K., Salazarm A.M., Dannenberg, A.L., Ommaya, A.K., Chervinsky, A.B., & Schwab, K. (1996). Outcome after traumatic brain injury in the U.S. military medical system. *The Journal of Trauma*, **41**, 972-975.

Raine, A. (1993). *The psychopathology of crime*. San Diego: Academic Press.

Raine, A., Brennan, P., & Mednick, S.A. (1994). Birth complications combined with early maternal rejection at age 1 year predispose to violent crime at age 18. *Archives of General Psychiatry*, **51**, 984-988.

Raine, A., Buchsbaum, M.S., & La Casse, L. (1997). Brain abnormalities in murderers indicated by positron emission tomography. *Biological Psychiatry*, **42**, 495-508.

Raine, A., Buchsbaum, M.S., Stanley, J., Lottenberg, S., Abel, L., & Stoddard, J. (1994). Selective reductions in pre-frontal glucose metabolism in murderers. *Biological Psychiatry*, **36**, 365-373.

Raine, A., & Venables, P.H. (1988). Enhanced P3 evoked potentials during a continuous performance task in psychopaths. *Psychophysiology*, **25**, 30-38.

Raine, A., Venables, P.H., & Williams, M. (1996). Better autonomic conditioning and faster electrodermal half-recovery time at age 15 years as possible protective factors against crime at age 29 years. *Developmental Psychology*, **32**, 624-630.

Richardson, J.T.E. (2000). *Clinical and neuropsychological aspects of closed head injury*. Hove: Psychology Press.

Roberts, A.C., Robbins, T.W., & Weiskrantz, L. (1998). *The prefrontal cortex: Executive and cognitive functions*. Oxford: Oxford University Press.

Sarapata, M., Herrmann, D., Johnson, T., & Aycock, R. (1996). The role of head injury in cognitive functioning, emotional adjustment and criminal behaviour. *Brain Injury*, **10**, 821-842.

Spellacy, F. (1977). Neuropsychological differences between violent and non-violent adolescents. *Journal of Clinical Psychology*, **33**, 966-969.

Taylor, P.J., Brown, R., & Gunn, J. (1983). Violence, psychosis and handedness. In P. Flor-Henry & J. Gruzelier (Eds.), *Laterality and psychopathology* (pp. 181-194). Amsterdam: Elsevier

Teasdale, G., & Mendelow, D. (1981). Pathophysiology of head injury. In N. Brooks (Ed.), *Closed head injury* (pp. 4-36). Oxford: Oxford University Press.

Walsh, K.W., & Darby, D. (1999). *Neuropsychology: A clinical approach* (4th ed.) Edinburgh: Churchill Livingstone.

Wood, R.L. (1984). Behaviour disorders following severe brain injury: their presentation and psychological management. In N. Brooks (Ed.), *Closed head injury* (pp. 195-219). Oxford: Oxford University Press.

PART D

NEUROCOGNITIVE MODELS AND PSYCHOBIOLOGICAL MARKERS FOR CRIMINALITY

Chapter 7

NEURO-COGNITIVE MODELS OF ACQUIRED SOCIOPATHY AND DEVELOPMENTAL PSYCHOPATHY

R. James R. Blair, Ph.D.
Institute of Cognitive Neuroscience and Department of Psychology, University College London

Acknowledgments. This work was supported by the Medical Research Council [ref. G9716841] and Department of Health [VISPED initiative].

INTRODUCTION

The goal of this chapter is to consider neuro-cognitive models of "acquired sociopathy" and developmental psychopathy. I will first provide definitions of these two clinical conditions. I will then detail the differences in the form of aggression—reactive and instrumental—that these two disorders present with. Following this, I will review what is known about the neural bases of these disorders on the basis of neuro-imaging studies. Then, two models of acquired sociopathy will be described: the somatic marker hypothesis and the social response reversal model. I will also consider whether either of these models could account for developmental psychopathy. Finally, I will consider a model of developmental psychopathy that has been extremely successful in accounting for much of the data on this disorder.

DEFINING "ACQUIRED SOCIOPATHY" AND DEVELOPMENTAL PSYCHOPATHY

"Acquired sociopathy" was a term introduced by Damasio (e.g., 1994) to characterize individuals who, following acquired lesions of orbitofrontal cortex, fulfill the DSM-III diagnostic criteria for "sociopathic disorder" (American Psychiatric Association, 1980). Previously, Blumer and Benson had used the term "pseudo-psychopathy" to refer to patients with frontal lobe lesions presenting in this manner (Blumer & Benson, 1975). The diagnostic criteria of "sociopathic disorder" that Damasio and colleagues believed their patients displayed included "reckless regarding others' personal safety," "lack of remorse," "inability to sustain consistent work behavior" and "defective planning" (Damasio, Tranel, & Damasio, 1990). Damasio and colleagues' most well known patient, EVR, was a successful, happily married professional with two children, prior to his orbitofrontal meningioma. However, subsequent to his surgery, his social conduct was profoundly affected; he was divorced twice and entered into disastrous business ventures that led to predictable bankruptcy.

Developmental psychopathy is a profoundly disturbing disorder that is characterized by persistent criminality and a marked emotional impairment, consisting of impoverished empathy, lack of remorse, and a predatory interpersonal style (Cleckley, 1967; Hare, 1991). It is identified according to a standard checklist of behaviors: for children, the Psychopathy Screening Device (PSD: Frick & Hare, in press) and for adults, the Revised Psychopathy Checklist (PCL-R: Hare, 1991). Importantly, this disorder is not equivalent to the psychiatric diagnoses of Conduct Disorder or Antisocial Personality Disorder (American Psychiatric Association, 1994). These psychiatric diagnoses concentrate almost entirely on the antisocial behavior shown by the individual rather than on any form of functional impairment. This has resulted in rates of diagnosis of Conduct Disorder reaching up to 16% of boys in mainstream education (American Psychiatric Association, 1994) while rates of diagnosis of Antisocial Personality Disorder are over 80% in forensic institutions (Hart & Hare, 1996). Because of these high rates of diagnosis, populations identified with these diagnostic tools can be highly heterogeneous. Psychopathy, in contrast, is shown by less than 1% of individuals in mainstream education (Blair & Coles, 2000) and less than 30% of individuals incarcerated in forensic institutions (Hart & Hare, 1996).

Clear differences are beginning to emerge between the conditions of "acquired sociopathy" and developmental psychopathy, not least that "acquired sociopathy" is observed following acquired lesions of orbitofrontal cortex, while psychopathy is a developmental disorder that appears at an early age and continues throughout the lifespan (see Harpur & Hare, 1994). However, while the previous distinction could just represent differences in cause of a functionally similar disorder, there is one difference that is of theoretical significance. Patients with "acquired sociopathy," if they present with aggression, present with *reactive* aggression (Anderson, Bechara, Damasio, Tranel, & Damasio, 1999; Blair & Cipolotti, 2000; Grafman, Schwab, Warden, Pridgen, & Brown, 1996). Their aggression is sporadic, stress-related and more frequently directed against physical objects than other individuals. As Burgess and Wood (1990) noted, "following serious [frontal] brain injury many patients find it difficult to tolerate frustration and react impulsively to minor forms of pressure or provocation" (p. 122). In contrast, individuals with developmental psychopathy frequently show high levels of *repetitive* aggression, significantly higher than non-psychopathic inmates (Hare & Jutai, 1983). Moreover, they present with a marked preponderance for goal-directed instrumental aggression rather than reactive aggression (Cornell *et al.*, 1996; Williamson, Hare, & Wong, 1987).

REACTIVE AND INSTRUMENTAL AGGRESSION

A considerable amount of theoretical and empirical research has made a distinction between reactive and instrumental antisocial behavior (e.g., Barratt, Stanford, Dowdy, Liebman, & Kent, 1999; Berkowitz, 1974, 1993; Crick & Dodge, 1996; Linnoila *et al.*, 1983). Reactive, impulsive aggression generally refers to unplanned aggressive acts which are spontaneous in nature, are either unprovoked or out of proportion to the provocation, and occur among persons who are often characterized as 'having a short fuse.' Reactive aggression is frequently elicited in response to frustration/threat. Instrumental, non-impulsive aggression is purposeful and goal directed and can be described as 'cold-blooded.' Reactive and instrumental aggression have been shown to be independent constructs in both children and adults through interviews and laboratory tasks (Barratt, Stanford, Kent, & Felthous, 1997).

Reactive and instrumental aggression are probably mediated by different neural architectures. Reactive aggression can be plausibly related to the eventual fight response of an animal approached by a threat. Most animals show a graded response to an approaching threat: they will freeze, then attempt to escape and, if this is impossible, attack the threat (Blanchard, Blanchard, & Takahashi, 1977). This graded response to an approaching threat is primarily mediated by subcortical systems (Panksepp, 1998; Siegel, Roeling, Gregg, & Kruk, 1999). It is plausible that the reactive aggression shown by some patients with "acquired sociopathy" occurs because of reduced control of these subcortical systems. In contrast, instrumental aggression is goal-directed motor behavior, and is likely to recruit the same cortical neural systems as any other goal-directed motor program, such as reaching for an object. In brief, these neural systems would include temporal cortex, to represent the object, and striatal and premotor cortical neurons to implement the actual behavior. The amygdala and ventral orbitofrontal cortex are crucially involved in learning that a particular behavior is the right one to achieve a goal (e.g., see Killcross, Robbins, & Everitt, 1997; Murray, Bussey, & Wise, 2000; Passingham & Toni, 2001; Passingham, Toni, & Rushworth, 2000). Thus, if the amygdala provides a reward signal following the instrumental aggression, the individual will be more likely to be instrumentally aggressive in the future. Conversely, if it provides a punishment signal, that person will be less likely to be instrumentally aggressive in the future.

NEURAL SYSTEMS IMPLICATED IN "ACQUIRED SOCIOPATHY" AND DEVELOPMENTAL PSYCHOPATHY: BRAIN-IMAGING DATA

Frontal-lobe damage has long been linked to emotional and personality changes such as euphoria, irresponsibility, lack of affect, and lack of concern for the present or future (e.g., Hecaen & Albert, 1978; Stuss & Benson, 1986). Frequently, increased levels of reactive aggression are reported (e.g., Burgess & Wood, 1990; Grafman *et al.*, 1996; Pennington & Bennetto, 1993). Analysis of the lesions of the patients affected has allowed some increase in specification regarding which areas of frontal cortex are particularly implicated in the control of reactive aggression. These are orbito- (ventral) and medial-frontal cortex (Damasio, 1994; Grafman *et al.*, 1996; Volavka, 1995).

At present, no direct data obtained through brain imaging are available on the neural systems implicated in developmental psychopathy. However, two functional (Intrator *et al.*, 1997; Schneider *et al.*, 2000) and two structural (Laakso *et al.*, 2001; Raine, Lencz, Bihrle, LaCasse, & Colletti, 2000) imaging studies have been conducted on individuals scoring highly on the PCL-R (though, it should be noted that the majority of the experimental populations in all four of these studies did not score more than the 30 point cut off to meet criteria for psychopathy). One functional study using Single Photon Emission Computerized Tomography (SPECT) and a regions-of-interest approach, investigated the neural response to emotional and neutral words in individuals scoring highly on the PCL-R (Intrator *et al.*, 1997). The second functional study used functional Magnetic Resonance Imaging (fMRI) to explore the neural response during conditioned aversion in individuals scoring highly on the PCL-R (Schneider *et al.*, 2000). An additional two studies have used MRI to explore structural differences in individuals scoring highly on the PCL-R (Laakso *et al.*, 2001; Raine *et al.*, 2000).

The Intrator *et al.* (1997) study found that within a mid-ventricular slice, the high PCL-R scoring individuals showed *greater* frontotemporal activation to emotional words relative to neutral words than did the low PCL-R scoring individuals. Similarly, the Schneider *et al.* (2000) study observed *greater* activation to conditioned stimuli in dorsolateral prefrontal cortex and amygdala in the high PCL-R scoring individuals relative to the comparison individuals. However, it should be noted that the Schneider *et al.* study also found superior aversive conditioning in this population, which is inconsistent with previous research (see below). Psychopathic individuals are notable for their poor emotional learning (Lykken, 1957; Newman & Kosson, 1986; Newman, Patterson, & Kosson, 1987). Moreover, the anatomical findings in the individuals scoring highly on the PCL-R were very similar to those obtained earlier by the same group with a population of individuals presenting with social phobia (Schneider *et al.*, 1999). Thus, these results should be considered tentative.

The Laakso *et al.* (2001) study investigated hippocampal volumes in individuals scoring highly on the PCL-R and a comparison population. This study reported negative correlations between psychopathy scores and the posterior half of the hippocampi bilaterally. The Raine *et al.* (2000) study investigated prefrontal white and gray matter volumes in individuals scoring highly on the PCL-R and two comparison populations. This study reported reduced prefrontal gray, but not white, matter volume in the individuals scoring highly on the PCL-R.

In addition to these two studies on populations scoring highly on the PCL-R, there have been several brain-imaging studies of aggressive individuals (Critchley *et al.*, 2000; Goyer *et al.*, 1994; Raine *et al.*, 1994; Raine, Buchsbaum, & LaCasse, 1997; Raine, Phil, Stoddard, Bihrle, & Buchsbaum, 1998; Schneider *et al.*, 2000; Volkow & Tancredi, 1987; Volkow *et al.*, 1995; Wong *et al.*, 1997). Volkow and Tancredi (1987) examined cerebral blood flow (CBF) under rest conditions using Positron Emission Tomography (PET) in four reactively violent psychiatric patients. All four presented with reduced CBF in the left temporal lobe and two also presented with reduced CBF in frontal cortex. In a follow-up study, Volkow *et al.* (1995) examined CBF under rest conditions using PET in eight reactively violent psychiatric patients and eight comparison individuals. As a group, the violent individuals showed significantly less CBF in medial temporal and frontal cortex than the comparison individuals.

The Raine *et al.* (1994, 1997) studies examined CBF using PET during performance of a continuous performance task in murderers pleading not guilty by reason of insanity and matched comparison individuals (n = 22 & n = 41 in both groups in the 1994 and 1997 studies respectively). Raine *et al.* (1994) found reduced CBF in prefrontal cortex in the murderers pleading not guilty by reason of insanity. Raine *et al.* (1997) found reduced CBF in the murderers pleading not guilty by reason of insanity in the prefrontal cortex, superior parietal gyrus, left angular gyrus, and the corpus callosum, while abnormal asymmetries of activity (left hemisphere lower than right) were also found in the amygdala, thalamus, and medial temporal lobe. Raine *et al.* (1998) split their population of murderers into those with psychosocial deprivation (e.g., childhood abuse or family neglect) and those without. They found that only the population without deprivation showed reduced prefrontal functioning. It should be noted, however, that the Volkow and Tancredi (1987), and Raine *et al.* (1994, 1997, 1998) studies included violent individuals with known organic brain damage (this was the case for over 50% of the murderers in the Raine *et al.* studies) and in the studies conducted by Raine *et al.*, violent individuals with schizophrenia. Thus, the atypical blood flow reported may reflect organic damage or schizophrenia that may, or may not, be related their aggressive behavior. Interestingly, Wong *et al.* (1997) investigated CBF under rest conditions using PET, and structural abnormalities using MRI, in 20 repetitive violent schizophrenic offenders and a matched group of 19 non-repetitive violent schizophrenics. This study reported no group differences in functioning or structure.

The Goyer *et al.* (1994) study examined CBF under rest conditions using PET of 17 patients with personality disorder (antisocial, borderline, dependent, and narcissistic) and 43 comparison individuals. They compared CBF to their patients' history of (mostly reactive) aggression. Interestingly, they found lower normalized CBF in orbitofrontal cortex (BA 47) which was correlated with a history of aggression.

Critchley *et al.* (2000) investigated the neural integrity of prefrontal cortex and the amygdala-hippocampal complex using proton magnetic resonance spectroscopy (^{1}H-MRS) in violent patients with mild mental impairment. ^{1}H-MRS can be used to measure concentrations, and peak area ratios of N-acetyl aspartate (NAA), creatine/phosphocreatine (Cr1PCr) and choline-related compounds (Cho). NAA is considered to be a nonspecific marker of neuronal density (Christiansen, Toft, Larsson, Stubgaard, & Henriksen, 1993). Creatine and phosphocreatine form a single MRS peak, and are involved in high-energy phosphate storage and metabolism. A range of neural membrane constituents co-resonate to form a single peak (the Choline signal) which is highest from glial cells, and reflects membrane phospho-lipid metabolism (Urenjak, Williams, Gadian, & Noble, 1993). This study found that compared to controls, the violent patients had significantly lower prefrontal concentrations of NAA and Cr1PCr, and a lower ratio of NAA/Cr1PCr in the amygdala-hippocampal complex; that is, the violent patients with mild mental impairment presented with abnormalities in both areas investigated (there was no investigation of any area where abnormality was not predicted).

NEURO-COGNITIVE MODELS OF "ACQUIRED SOCIOPATHY" AND DEVELOPMENTAL PSYCHOPATHY

In conclusion, analysis of the locations of the lesions of patients with acquired sociopathy allows us to conclude that acquired sociopathy is associated with damage to orbitofrontal cortex and perhaps, medial frontal cortex. While there are no clear neuro-imaging results for individuals with developmental psychopathy, there are interesting indications that prefrontal cortex, orbitofrontal cortex and temporal lobe structures (including the amygdala and hippocampus) may be involved in increasing risk for aggression (see also Chapter 8 of this volume). In the following sections, I will consider two neurocognitive models of "acquired sociopathy": the somatic marker and social response reversal models. In addition, I will consider two models of psychopathy: the low fear and Violence Inhibition Mechanism models.

THE SOMATIC MARKER MODEL

According to Damasio and colleagues, ventromedial frontal cortex (orbitofrontal and medial frontal cortex) acts as a repository, and is involved in the formation of recorded dispositional linkages between factual knowledge and bio-regulatory states (Bechara, Tranel, & Damasio, 2000; Damasio, 1994). When individuals are faced with a situation for which some factual aspects have been previously categorized, the dispositional linkages are activated. This can occur via a 'body loop' in which a 'somatic marker' is conveyed to somato-sensory cortices, or via an 'as-if body loop' in which the body is bypassed and reaction signals are conveyed to the somato-sensory structures. Under either loop, the somato-sensory structures then adopt an appropriate pattern that constrains option-outcome reasoning. In short, the somato-sensory pattern marks the scenario as either good or bad, allowing the rapid rejection/endorsement of specific option-outcome pairs. With regards to reactive aggression within this model, we could speculate that an option-outcome pair is activated such as "hit that person but be punished later." In a healthy individual, there will be activation of the linkage between knowledge of hitting and punishment and the emotional aversion to punishment. The consequent aversive somatic marker should then guide the individual away from hitting the other. However, if there is damage to the somatic marker system, there will be no somatic marker to guide behavior and the individual may hit the person.

There are two main sets of data that are associated with the somatic marker hypothesis. First, patients with lesions to ventromedial frontal cortex fail to show autonomic responses to visually presented social stimuli (scenes of social disaster, mutilation and nudity) under passive viewing conditions (Damasio *et al.*, 1990; Damasio, Tranel, & Damasio, 1991; Tranel & Damasio, 1994). Passive viewing conditions require the participant to only look at the visual stimuli. It should be noted that these patients did, however, show appropriate autonomic responses to the same stimuli under active viewing conditions. These involved attention-inducing instructions; for example, asking the participants to describe the pictures.

Secondly, patients with lesions to ventromedial frontal cortex perform poorly on the four-pack card-playing task (Bechara, Damasio, Damasio, & Lee, 1999; Bechara, Damasio, Tranel, & Damasio, 1997; Bechara, Damasio, Damasio, & Anderson, 1994; see, for a review, Bechara, Damasio, & Damasio, 2000). In this task, participants are presented with four packs of cards. Two packs result in high rewards but even higher punishments and, if played continuously, result in a net loss. Two packs result in low rewards but even lower punishments and, if played continuously, result in a net gain. The participants have to learn to avoid the high reward, net loss packs in favour of the low reward, net gain packs. Healthy participants learnt to take from the low reward packs and showed skin conductance responses (i.e., warning somatic markers) before the selection of a card from the disadvantageous packs. In contrast, patients with ventromedial damage continued to choose from the disadvantageous packs and failed to show skin conductance responses before their choices from these packs.

Recently, Bechara *et al.* (2000) designed a variant of the original gambling task in which the advantageous decks yielded high immediate punishment but even higher future reward and the disadvantageous decks yielded low immediate punishment but even lower future reward. Patients with lesions to ventromedial frontal cortex opted for the disadvantageous decks in this variant of the gambling task also. These results, taken in conjunction with the results obtained with the previous variant, strongly suggest that patients with lesions to ventromedial frontal cortex are neither hypersensitive to reward nor insensitive to punishment. Bechara *et al.* interpret this data as suggesting that patients with lesions to ventromedial frontal cortex are insensitive to future consequences, positive or negative, because they lack the guiding influence of somatic markers.

Rogers and colleagues have developed a second decision-making task (Rahman, Sahakian, Hodges, Rogers, & Robbins, 1999; Rolls & Stringer, 2001; Rogers, Everitt, *et al.*, 1999). Performance on this task has been explained with reference to the somatic marker model (Rahman, Sahakian, Cardinal, Rogers, & Robbins, 2001). In this task, the participant has to state whether a yellow token is contained in either one of the red or one of the blue boxes presented to them. The ratio of red and blue boxes is manipulated from trial to trial, so a participant's decision-making behavior is examined over a variety of differentially weighted contingencies (e.g., 9 red to 1 blue as compared to 4 red to 6 blue). Following this choice, the participant is instructed to place a "bet" on his or her choice being correct. The available bets appear in a sequence, one after another, each for 5 seconds, and the participant can select any of these bets. Following their selection, the participant is informed whether he or she is correct or incorrect and either wins or loses the points that he/she bet. The betting occurs in both ascending (the first bet offered is small but replaced by a larger amount) and descending (the first bet offered is large but replaced with a smaller amount) conditions. The participant's choice of contingency, speed of choice, and size of bet are recorded.

This decision-making task is complex and its computational underpinnings remain unclear. However, it is notable that patients with orbitofrontal cortex lesions show increased deliberation times, poorer decision making (as indexed by their greater rates of picking the less likely outcome) and bet less than comparison individuals, including patients with lesions to dorsolateral prefrontal cortex (Rogers, Everitt, *et al.*, 1999). Rogers *et al.* also conducted a PET study of a variant of the task, where the healthy participant had to predict which of two mutually exclusive outcomes would occur where, critically, the larger reward (and penalty) was associated with choice of the least likely outcome, whereas the smallest reward (and penalty) was associated with choice of the most likely outcome. Resolving these "conflicting" decisions was associated with three distinct foci of regional cerebral blood flow increase within the right inferior and orbital prefrontal cortex (Rogers, Owen, *et al.*, 1999).

However, while it should be noted that the four-pack card-playing task and the decision-making task may characterize the decision-making impairments shown by patients with orbitofrontal cortex lesions, it is less clear that they characterize the reasons for the reactive aggression shown by some patients with orbitofrontal cortex lesions. Indeed, the descriptions of most of the patients in these studies do not indicate individuals presenting with antisocial behavior. It could be argued that this is because reactive aggression and other forms of antisocial behavior are only seen in patients whose lesions occur early in life (Anderson *et al.*, 1999). However, there is now considerable data suggesting that patients whose lesions are acquired in adulthood may also present with severe behavioral problems (Blair & Cipolotti, 2000; Burgess & Wood, 1990; Grafman *et al.*, 1996). Thus, while the somatic marker model may prove to be a useful model of decision making and impairment in decision making following orbitofrontal cortex lesions, there is less data to support it as a model of the control of reactive aggression.

It has also been suggested that early somatic-marker impairment could result in developmental psychopathy (Anderson *et al.*, 1999; Damasio, 1994). However, there are several difficulties with this position. Firstly, while individuals with acquired sociopathy are generally autonomically hypo-responsive to visual stimuli (at least under passive viewing conditions), psychopathic individuals show selective hypo-responsiveness to sad and fearful expressions (Aniskiewicz, 1979; Blair, Jones, Clark, & Smith, 1997; House & Milligan, 1976). Secondly, while individuals with acquired sociopathy show intact autonomic aversive conditioning (Bechara *et al.*, 1999), psychopathic individuals present with impaired autonomic aversive conditioning (Lykken, 1957). Thirdly, individuals with acquired sociopathy present with reactive aggression (Anderson *et al.*, 1999; Blair & Cipolotti, 2000; Grafman *et al.*, 1996). In contrast, psychopathic individuals present with a marked preponderance for goal-directed instrumental aggression (Cornell *et al.*, 1996; Williamson *et al.*, 1987). The data suggest that the somatic marker position cannot provide an explanation for the instrumental aggression associated with developmental psychopathy.

THE SOCIAL RESPONSE REVERSAL MODEL

An alternative account of "acquired sociopathy" focusses on explaining the appearance of reactive aggression rather than on providing an account of the decision-making impairment shown by this patient group. This position makes reference to the role of orbitofrontal cortex in response reversal as a consequence of changes in reinforcement contingencies (Dias, Robbins, & Roberts, 1996; Rolls, 1997). However, this position stresses the role of social cues in modulating social behavior (Blair, 2001; Blair & Cipolotti, 2000). Angry expressions are known to curtail the behavior of others in situations where social rules or expectations have been violated (Averill, 1982). It is suggested that there is a system that is activated by another's angry expressions and perhaps by representations of situations that have been previously associated with another individual's angry responses or other negative valence expressions (e.g., the staring expressions of others that can precede a sense of embarrassment and perhaps others' disgusted expressions); that is, situations where another individual's anger might be expected. The suggestion is that activation of this system results in the modulation of current behavioral responding, in particular the modulation of reactive aggression mediated by the hypothalamus and periaqueductal gray (and perhaps sexual behavior mediated by related systems). Evolutionarily, this system may have its origins in regulating behavior in disputes between conspecifics at different levels in a social hierarchy. Decisions concerning whether to fight/flight conspecifics are heavily influenced by such cues as the other conspecifics' hierarchical location and whether they are currently displaying threat cues (de Waal, 1992).

This hypothesis has drawn support from findings indicating that orbitofrontal cortex (Brodmann's Area 47) is activated by negative emotional expressions; in particular, anger but also fear and disgust (Blair, Morris, Frith, Perrett, & Dolan, 1999; Kesler/West *et al.*, 2001; Sprengelmeyer, Rausch, Eysel, & Przuntek, 1998). In addition, this region is also activated under conditions when an individual is induced to feel angry (Dougherty *et al.*, 1999). Moreover, patients with social behavioral difficulties following lesions of frontal cortex, particularly the orbitofrontal region, are also impaired in the recognition of facial expressions, particularly anger (Blair & Cipolotti, 2000; Hornak, Rolls, & Wade, 1996). In addition, both alcohol and diazepam have been found to selectively impair the ability of healthy individuals to process angry expressions (Blair & Curran, 1999; Borrill, Rosen, & Summerfield, 1987) and both are associated with increased risk for reactive aggression (Bond, Curran, Bruce, O'Sullivan, & Shine, 1995).

Patients with social behavioral difficulties following lesions of frontal cortex, particularly the orbitofrontal region, are not only impaired in the recognition of angry facial expressions. Such patients have also been found to show impairment in appropriately attributing anger and embarrassment—but not other emotions—to story protagonists (Blair & Cipolotti, 2000). In addition, such patients have a deficit in identifying the sorts of violations of social norms that induce anger in others (Blair & Cipolotti, 2000; Stone, Baron-Cohen, & Knight, 1998). Interestingly, a recent fMRI study revealed activation of the region of orbitofrontal cortex (BA 47) that responds to angry facial expressions when individuals were asked to read social norm violations (Berthoz, Armony, Blair, & Dolan, submitted).

The main difference between the Social Response Reversal (SRR) and Somatic Marker positions is in the set of stimuli thought to activate both systems. For SRR, this is primarily angry expectations or the expectation of another's anger. In contrast, the somatic marker system is thought to be activated by a broad range of stimuli. Thus, in contrast to the somatic marker position, the SRR position predicts that generating judgments on the basis of an expectation of another's anger should be dissociable from altering behavior as a function of point rewards/punishments on, for example, the four-pack card-playing task. Moreover, this prediction appears to be upheld. Patients with "acquired sociopathy" have been observed with SRR impairment, but show preserved performance on the four-pack card-playing task (Blair & Cipolotti, 2000). In addition, adult developmental psychopaths show no SRR impairment, but do show impairment on response-reversal tasks (Blair & Cipolotti, 2000; LaPierre, Braun, & Hodgins, 1995).

A NEURO-COGNITIVE MODEL OF DEVELOPMENTAL PSYCHOPATHY

In the sections above, I have described two models of "acquired sociopathy", the Somatic Marker and SRR models. While these models provide interesting accounts of the decision making and social impairments shown by patients with acquired sociopathy, neither can be considered as potential accounts of developmental psychopathy. Both the somatic marker and SRR positions predict impairments that are not seen in psychopathic individuals (see above). I now turn to a model that can account for a considerable amount of the data on psychopathic individuals (Blair, 1995, 2001).

The clinical and empirical picture of the psychopath describes an individual who has some form of emotional deficit (Aniskiewicz, 1979; Cleckley, 1967; Hare, 1991; Patrick, 1994; Patrick, Cuthbert, & Lang, 1994; Schalling, 1978; Williamson, Harpur, & Hare, 1991). One of the main forms of emotional impairment shown by psychopathic individuals is in fear-related activity (Fowles, 1988; Hare, 1978; Lykken, 1995; Mealey, 1995; Patrick, 1994). The lack of emotional responsiveness of psychopaths to threatening stimuli has been demonstrated through several paradigms (Hare, 1982; Levenston, Patrick, Bradley, & Lang, 2000; Lykken, 1957; Ogloff & Wong, 1990; Patrick *et al.*, 1994). For example, psychopaths have been shown to be deficient in the acquisition of anxiety responses to threatening stimuli (e.g., Lykken, 1957). They have also been found, when anticipating aversive shock, to show smaller electrodermal responses that occur later in the warning interval in comparison to non-psychopaths (Hare, 1965, 1982; Hare, Frazelle, & Cox, 1978; Ogloff & Wong, 1990). In addition, psychopathic individuals show reduced autonomic reactivity relative to non-psychopathic controls, during the imagery of unpleasant and fearful experiences (Patrick *et al.*, 1994). Only if the threatening stimulus is a picture (e.g., of a snake), have psychopaths been found to show similar responses to non-psychopaths (Levenston *et al.*, 2000; Patrick, Bradley, & Lang, 1993). However, this finding has only been found to hold if skin conductance response indexes responsiveness. If it is indexed by startle reflex, psychopaths have been found to show reduced responsiveness to these stimuli relative to controls (Levenston *et al.*, 2000). For an extensive overview of such psychophysiological research, see Chapter 8.

A second form of emotional impairment shown by psychopathic individuals is in empathic responding (Blair, 1995; Hare, 1991). The importance of empathy for moral socialization was one of the reasons for the development of the original Violence Inhibition Mechanism (VIM) model of Psychopathy (Blair, 1995; Blair *et al.*, 1997). This model was prompted by ethological work suggesting that most social animals possess mechanisms for the control of aggression (Eibl-Eibesfeldt, 1970; Lorenz, 1966). The ethologists noted that submission cues displayed to a conspecific aggressor terminate attacks; for example, an aggressor dog will cease fighting if its opponent bares its throat. The VIM is considered to be a functionally similar mechanism in humans, where sad and fearful facial affects (i.e., distress cues) serve as a human submission response. The suggestion is that the VIM is activated by distress cues, resulting in increased autonomic activity, attention and activation of the brainstem threat response system (usually resulting in freezing; Blair, 1995). The suggestion is that the VIM is activated whenever distress cues are displayed. It is not reliant upon contextual information about ongoing violence for activation. In line with this, the display of distress cues has been found to result in the inhibition of not only aggression (Perry & Perry, 1974) but also non-violent disputes over property ownership (Camras, 1977) and sexual activity (Chaplin, Rice, & Harris, 1995).

One early index of appropriate moral socialization, and thus the developmental integrity of the Violence Inhibition Mechanism, is the demonstration by the child of the moral/conventional distinction. From the age of 3.5 years, children distinguish in their judgments between moral (victim-based) and conventional (social disorder-based) transgressions (see Smetana, 1993). Crucially for the model (Blair, 1995), normally developing children best discriminate in their judgments between two types of transgressions when they are asked to imagine situations where there are no rules prohibiting the transgressions. In contrast, adults with psychopathy and children with psychopathic tendencies are least likely to make a discrimination under these conditions (Blair, 1995, 1997; Blair, Jones, Clark, & Smith, 1995; see also Nucci & Herman, 1982 and Arsenio & Fleiss, 1996 for similar work with children with Behavior Disorder and Conduct Disorder). In addition, and in line with the VIM position, psychopathic adults show reduced comprehension of situations likely to induce guilt, although they do show appropriate comprehension of happiness, sadness and even complex emotions such as embarrassment

(Blair, Sellars *et al.*, 1995). Finally, and directly predicted by the model, psychopathic adults and children with psychopathic tendencies show reduced autonomic activity to the sadness and fear of others (Aniskiewicz, 1979; Blair, 1999; Blair *et al.*, 1997; House & Milligan, 1976).

The VIM model has been expanded at both the cognitive and neural levels (Blair, 2001, in press; see Figure 1). The first difference between this new model and the previous formulation of the VIM position is that two trigger data bases are specified. The second difference is that the new model makes more reference to a putative anatomical substrate. Following the work of Killcross *et al.* (1997), a distinction is made between a trigger base for reflexive, automatic responses to conditioned stimuli and a trigger base for instrumental behavior based on emotional events. It is this second trigger base that allows the activation of motor programs. Again following Killcross *et al.*, these trigger databases are attributed to the central and basolateral nuclei of the amygdala respectively. Through association, pairings of the activation of representations of conditioned stimuli (CSs) held in temporal cortex with activation by the unconditioned stimuli (USs) of aversive and appetitive unconditioned stimuli (including sad faces) will result in the CSs activating the autonomic nervous system and a basic threat response system mediated by the brainstem. In addition, reinforcement/punishment of actions from the trigger base for instrumental behavior will increase/decrease the probability of these actions occurring in the future. The appropriately developing child thus initially finds the pain of others' aversive and then, through socialization, finds thoughts of acts that cause pain to others aversive also. It is proposed that psychopathic individuals have had disruption to this system, and so the formation of these emotional associations is dysfunctional. The disruption is thought to be due to early amygdala dysfunction

The suggestion that psychopathy is due to amygdala dysfunction predicts that psychopathic individuals should show impairment on tasks in which both the central and basolateral nucleus of the amygdala are implicated. Animal studies have implicated the central nucleus of the amygdala in aversive conditioning (Davis, 2000; Killcross *et al.*, 1997) and in the augmentation of the startle response by threat primes (Davis, 2000). Human patients with amygdala lesions show impaired aversive conditioning (Bechara *et al.*, 1995; LaBar, LeDoux, Spencer, & Phelps, 1995) and reduced augmentation of the startle reflex by threat primes (Angrilli *et al.*, 1996). In addition, psychopathic individuals also show impaired aversive conditioning (Lykken, 1957) and reduced augmentation of the startle reflex by threat primes (Levenston *et al.*, 2000; Patrick *et al.*, 1993). Moreover, in a recent neuro-imaging study, amygdala activation has been reported to stimuli that anticipated shock (though no shock was given; Phelps *et al.*, 2001). Psychopathic individuals have been found

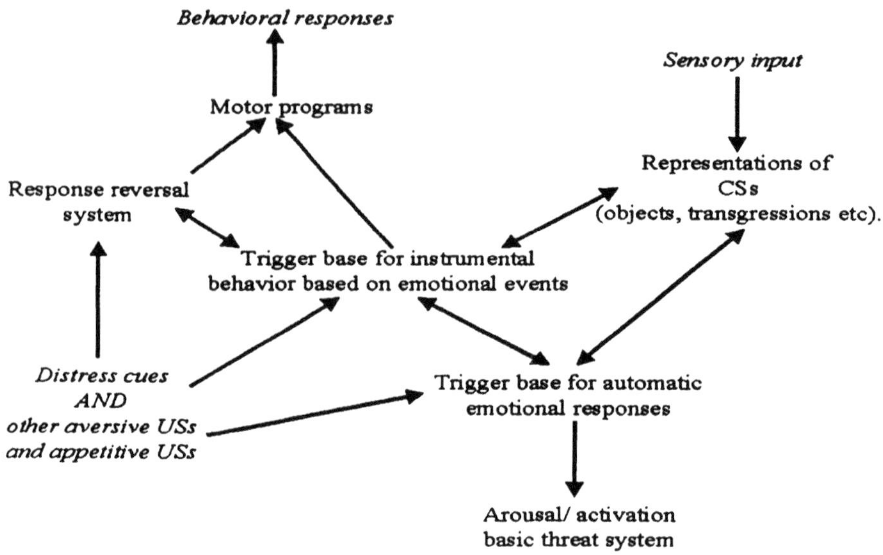

Figure 1: The revised Violence Inhibition Mechanism model

to show reduced electrodermal responses when anticipating shock (Hare, 1965, 1982; Hare *et al.*, 1978; Ogloff & Wong, 1990).

The position also predicts that psychopathic individuals should show impaired performance on tasks in which the basolateral nucleus of the amygdala is implicated. Thus, animal studies have implicated the basolateral nucleus of the amygdala in instrumental learning (Killcross *et al.*, 1997). Moreover, there is considerable work implicating the basolateral nucleus of the amygdala in passive avoidance learning (Ambrogi Lorenzini, Baldi, Bucherelli, Sacchetti, & Tassoni, 1999). There is preliminary work with human patients with amygdala lesions and psychopathic individuals showing impairments in instrumental learning (Fine, 2000). Moreover, psychopathic individuals show pronounced impairments on passive avoidance tasks (Newman & Kosson, 1986; Newman & Schmitt, 1998).

The role of the amygdala in processing the expressions of others is currently under debate. Thus, it has been suggested that the amygdala only responds to fearful expressions. Certainly, an impairment in the recognition of fearful expressions is the most common expression-processing abnormality found in patients with amygdala lesions (see, for a review, Fine & Blair, 2000; see also Adolphs *et al.*, 1999). In addition, functional imaging studies have shown that the amygdala is activated by fearful expressions (Morris *et al.*, 1996; Phillips *et al.*, 1997). However, this position has recently been challenged by Rapcsak *et al.* (2000), who reported no indications of disproportionate impairments for fearful expressions in their thirteen patients with amygdala lesions. It should be noted though that of Rapcsak *et al.*'s thirteen patients, eleven had had unilateral temporal lobectomies and the other two had had herpes encephalitis; that is, their lesions were extensive and included structures external to the amygdala. It is thus perhaps unsurprising that this study found little evidence of selective impairment. More challenging to the idea that the amygdala responds only to fearful expressions are neuropsychological findings indicating that patients with amygdala lesions are frequently impaired for the recognition of sad and angry expressions (see Fine & Blair, 2000), and neuro-imaging findings that sad facial expressions activate the amygdala (Blair *et al.*, 1999; Schneider *et al.*, 1995). This strongly suggests that the amygdala does not only respond to fearful facial expressions.

An alternative position is that the amygdala is part of a general neural system for recognizing emotions that signals potential harm to the organism; that is, fear and anger (Adolphs *et al.*, 1999). The major problem for this position is that reports of impairments in processing sad expressions in patients with amygdala lesions are at least as common as reports of patients with amygdala lesions being impaired in processing angry expressions (Fine & Blair, 2000). Moreover, and crucially, angry expressions have not been found to activate the amygdala in neuro-imaging studies (Blair *et al.*, 1999; Sprengelmeyer *et al.*, 1998; Whalen, Shin, McInerney, & Rauch, 1998).

An alternative position, and the one favored here, is developed from the work on the amygdala's role in emotional associative learning (Davis, 2000; Everitt, Cardinal, Hall, Parkinson, & Robbins, 2000; LeDoux, 2000). The suggestion is that the amygdala is crucially involved in processing those expressions that act as social unconditioned stimuli; that is, fearful and sad expressions (Blair *et al.*, 1999). In line with this position, both fearful and sad expressions have been found to activate the amygdala in neuro-imaging studies (Blair *et al.*, 1999; Morris *et al.*, 1996; Phillips *et al.*, 1997; Schneider *et al.*, 1995). However, it should be noted that patients with amygdala lesions frequently do not show impairment in processing sad expressions. Indeed, in a recent review, only 6/13 patients with amygdala lesions presented with impairment for sad expressions, while 10/13 presented with impairment for fearful expressions (Fine & Blair, 2000). The fact that fewer patients with amygdala lesions show impairment for the recognition of sad facial expressions may be due to the relative ease, compared to fearful expressions, with which sad expressions can be recognized (Ekman & Friesen, 1976).

As regards psychopathy, children with psychopathic tendencies show impairment when processing sadness and fearfulness, but not when processing expressions of anger, disgust, happiness or surprise (Blair & Coles, 2000; Blair, Colledge, Murray, & Mitchell, 2001; Stevens, Charman, & Blair, 2001). Interestingly, adult psychopathic individuals show a pattern of impairment more similar to that of patients with amygdala lesions. For adult psychopathic individuals, recognition of sad expressions appears to be too easy, and thus while they may show reduced autonomic activity to this expression (Aniskiewicz, 1979; Blair, 1999; Blair *et al.*, 1997; House & Milligan, 1976), they only show an expression recognition impairment for fearful expressions (Blair, Mitchell, & Colledge, submitted). In addition, children and adults with psychopathy show selective impairments when processing sad and fearful affect in vocal tones (Blair *et al.*, submitted). Neuro-imaging results have implicated the amygdala in the response to, at least, fearful vocal affect (Morris, Scott, & Dolan, 1999; Phillips *et al.*, 1998) although the neuropsychological literature has been less consistent (Anderson & Phelps, 1998; Scott *et al.*, 1997).

In Figure 1, the connections from the trigger bases to the representations of the conditioned stimuli are shown as being reciprocal, in line with suggestions that amygdala activation strengthens the activity of the representations that have activated the amygdala; that is, increasing "attention" to these representations. Thus, the activation of the representations of emotional words (e.g., murder) may be boosted by the reciprocal interactions with the amygdala if these words have become conditioned stimuli for amygdala activity. In lexical decision tasks, where the participant has to decide whether a letter string is a word or not, participants are faster to conclude that emotional words are words than to respond that neutral words are words (Graves, Landis, & Goodglass, 1981; Strauss, 1983). This is potentially because of the emotional word's increased salience due to activation from the amygdala. Interestingly, psychopathic individuals fail to show any reaction time differential between neutral and emotional words (Williamson *et al.*, 1991).

A final difference between this new model and the previous formulation of the VIM position is that connections with other systems are considered; in particular, systems allowing response reversal following changes in contingency. The interconnections between the basolateral nucleus of the amygdala and orbitofrontal cortex have a critical role in encoding and using associative information about the motivational significance of stimuli (Gallagher, McMahan, & Schoenbaum, 1999; Schoenbaum, Chiba, & Gallagher, 1998, 2000). Within this circuit, the amygdala has a critical role in forming associations between unconditioned and conditioned stimuli and between conditioned stimuli (Davis, 2000; Killcross *et al.*, 1997; Schoenbaum *et al.*, 1998). Orbitofrontal cortex encodes the motivational significance of the cues and the incentive value of expected outcomes (Gallagher *et al.*, 1999; Schoenbaum *et al.*, 2000; Thorpe, Rolls, & Maddison, 1983). Orbitofrontal cortex acts as a system for response reversal—when the incentive value of an expected outcome differs from the actual outcome, orbitofrontal cortex is involved in altering behavior. But both the basolateral amygdala and orbitofrontal cortex are involved in encoding the value of a stimulus, and lesions to either system impair reversals in responding following changes in the value of reinforcement (Gallagher *et al.*, 1999; Hatfield, Han, Conley, Gallagher, & Holland, 1996; Rolls, 1997). Indeed, both amygdala and orbitofrontal cortex dysfunction have been found to result in poor performance on the gambling task (Bechara *et al.*, 1999). In line with the suggestion of amygdala dysfunction in psychopathic individuals, both children with psychopathic tendencies and adult psychopaths show impairment on this task (Blair, Mitchell & Colledge, in press; Mitchell *et al.*, submitted). Failure in this system may also be the reason that children with psychopathic tendencies and adult psychopathic individuals continue to play cards on the one pack card playing task despite the escalating increase in punishment levels (Fisher & Blair, 1998; Newman *et al.*, 1987; O'Brien & Frick, 1996; Shapiro, Quay, Hogan, & Schwartz, 1988).

It is interesting to note, however, that the impairment shown by adult psychopathic individuals on Bechara's gambling task is considerably greater than that shown by children with psychopathic tendencies. Moreover, on simpler response reversal tasks, where the contingencies change from always rewarded for the response to always punished, adult psychopathic individuals show impairment (LaPierre *et al.*, 1995; Mitchell *et al.*, submitted), whereas children with psychopathic tendencies do not (Blair, Mitchell, & Colledge, in press). This suggests that the impairment in this system is more pronounced in adult psychopaths than in children with psychopathic tendencies. Children with psychopathic tendencies may be able to weakly represent the incentive value of expected outcomes and thus fail only on those tasks where the contingency change is very subtle. In contrast, adult psychopaths may have very limited ability to represent the incentive value of expected outcomes, and thus fail even those tasks where the contingency change is abrupt. Why should there be this difference between children with psychopathic tendencies and adult psychopaths? It may reflect gradual degeneration of the circuit connecting the amygdala and orbitofrontal cortex across development, as documented by Bachevalier (1994). Alternatively, it may reflect the lifestyle of adult psychopathic individuals, for whom substance abuse is very common (Hare, 1991). Drug abuse has been associated with orbitofrontal cortex dysfunction (Rogers & Robbins, 2001).

CONCLUSIONS

In this chapter I have described two models of "acquired sociopathy." I concluded that the Somatic Marker position may explain the decision-making impairments shown by patients with acquired sociopathy, but that the Social Response Reversal position more easily accounted for the reactive aggression that some of the patients present with. Following this, I developed a neuro-cognitive model of psychopathy that is capable of integrating the data obtained from the response set modulation, fear positions and Violence Inhibition Mechanism model. This model assumes a fundamental deficit within the amygdala that is responsible for the individual's failure to be concerned by his/her victims and their lack of moral socialization. In addition, it is assumed that during development increased dysfunction may be seen in related systems.

REFERENCES

Adolphs, R., Tranel, D., Young, A.W., Calder, A.J., Phelps, E.A., Anderson, A.K., Lee, G.P., & Damasio, A.R. (1999). Recognition of facial emotion in nine individuals with bilateral amygdala damage. *Neuropsychologia*, **37**, 1111-1117.

Ambrogi Lorenzini, C.G., Baldi, E., Bucherelli, C., Sacchetti, B., & Tassoni, G. (1999). Neural topography and chronology of memory consolidation: A review of functional inactivation findings. *Neurobiology of Learning and Memory*, **71**, 1-18.

American Psychiatric Association. (1980). *Diagnostic and statistical manual of mental disorders* (3rd ed., *DSM-III*). Washington, DC: American Psychiatric Association.

American Psychiatric Association. (1994). *Diagnostic and statistical manual of mental disorders* (4th ed.). Washington, DC: American Psychiatric Association.

Anderson, A.K., & Phelps, E.A. (1998). Intact recognition of vocal expressions of fear following bilateral lesions of the human amygdala. *Neuroreport*, **9**, 3607-3616.

Anderson, S.W., Bechara, A., Damasio, H., Tranel, D., & Damasio, A.R. (1999). Impairment of social and moral behavior related to early damage in human prefrontal cortex. *Nature Neuroscience*, **2**, 1032-1037.

Angrilli, A., Mauri, A., Palomba, D., Flor, H., Birhaumer, N., Sartori, G., & di Paola, F. (1996). Startle reflex and emotion modulation impairment after a right amygdala lesion. *Brain*, **119**, 1991-2000.

Aniskiewicz, A.S. (1979). Autonomic components of vicarious conditioning and psychopathy. *Journal of Clinical Psychology*, **35**, 60-67.

Arsenio, W.F., & Fleiss, K. (1996). Typical and behaviorally disruptive children's understanding of the emotion consequences of socio-moral events. *British Journal of Developmental Psychology*, **14**, 173-186.

Averill, J.R. (1982). *Anger and aggression: An essay on emotion*. New York: Springer-Verlag.

Bachevalier, J. (1994). Medial temporal lobe structures and autism: A review of clinical and experimental findings. *Neuropsychologia*, **32**, 627-648.

Barratt, E.S., Stanford, M.S., Dowdy, L., Liebman, M.J., & Kent, T.A. (1999). Impulsive and premeditated aggression: a factor analysis of self-reported acts. *Psychiatry Research*, **86**, 163-173.

Barratt, E.S., Stanford, M.S., Kent, T.A., & Felthous, A. (1997). Neuropsychological and cognitive psychophysiological substrates of impulsive aggression. *Biological Psychiatry*, **41**, 1045-1061.

Bechara, A., Damasio, A.R., Damasio, H., & Anderson, S.W. (1994). Insensitivity to future consequences following damage to human prefrontal cortex. *Cognition*, **50**, 7-15.

Bechara, A., Damasio, H., & Damasio, A.R. (2000). Emotion, decision making and the orbitofrontal cortex. *Cerebral Cortex*, **10**, 295-307.

Bechara, A., Damasio, H., Damasio, A.R., & Lee, G.P. (1999). Different contributions of the human amygdala and ventromedial prefrontal cortex to decision-making. *Journal of Neuroscience*, **19**, 5473-5481.

Bechara, A., Damasio, H., Tranel, D., & Damasio, A.R. (1997). Deciding advantageously before knowing the advantageous strategy. *Science*, **275**, 1293-1295.

Bechara, A., Tranel, D., & Damasio, H. (2000). Characterization of the decision-making deficit of patients with ventromedial prefrontal cortex lesions. *Brain*, **123**, 2189-202.

Bechara, A., Tranel, D., Damasio, H., Adolphs, R., Rockland, C., & Damasio, A.R. (1995). Double dissociation of conditioning and declarative knowledge relative to the amygdala and hippocampus in humans. *Science*, **269**, 1115-1118.

Berkowitz, L. (1974). Some determinants of impulsive aggression: role of mediated associations with reinforcements for aggression. *Psychological Review*, **81**, 165-176.

Berkowitz, L. (1993). *Aggression: Its causes, consequences, and control.* Philadelphia: Temple University Press.

Berthoz, S., Armony, J., Blair, R.J.R., & Dolan, R. (submitted). Neural correlates of violation of social norms and embarrassment.

Blair, R.J.R. (1995). A cognitive developmental approach to morality: Investigating the psychopath. *Cognition*, **57**, 1-29.

Blair, R.J.R. (1997). Moral reasoning in the child with psychopathic tendencies. *Personality and Individual Differences*, **22**, 731-739.

Blair, R.J.R. (1999). Responsiveness to distress cues in the child with psychopathic tendencies. *Personality and Individual Differences*, **27**, 135-145.

Blair, R.J.R. (2001). Neuro-cognitive models of aggression, the Antisocial Personality Disorders and Psychopathy. *Journal of Neurology, Neurosurgery & Psychiatry*, **71**, 1-4.

Blair, R.J.R. (in press). A neuro-cognitive model of the psychopathic individual. In T. Robbins, & M. Ron (Eds.), *Disorders of brain and mind II*. Cambridge: Cambridge University Press.

Blair, R.J.R., & Cipolotti, L. (2000). Impaired social response reversal: A case of "acquired sociopathy". *Brain*, **123**, 1122-1141.

Blair, R.J.R., & Coles, M. (2000). Expression recognition and behavioral problems in early adolescence. *Cognitive Development*, **15**, 421-434.

Blair, R.J.R., Colledge, E., Murray, L., & Mitchell, D. (2001). A selective impairment in the processing of sad and fearful expressions in boys with psychopathic tendencies. *Journal of Abnormal Child Psychology*, in press.

Blair, R.J.R., & Curran, H.V. (1999). Selective impairment in the recognition of anger induced by diazepam. *Psychopharmacology*, **147**, 335-338.

Blair, R.J.R., Jones, L., Clark, F., & Smith, M. (1995). Is the psychopath "morally insane"? *Personality and Individual Differences*, **19**, 741-752.

Blair, R.J.R., Jones, L., Clark, F., & Smith, M. (1997). The psychopathic individual: A lack of responsiveness to distress cues? *Psychophysiology*, **34**, 192-198.

Blair, R.J.R., Mitchell, D., & Colledge, E. (in press). Somatic markers and response reversal: Is there orbitofrontal cortex dysfunction in children with psychopathic tendencies? *Journal of Abnormal Child Psychology.*

Blair, R.J.R., Mitchell, D.G.V., & Colledge, E. (submitted). Reduced sensitivity to other's fearful expressions in psychopathic individuals.

Blair, R.J.R., Morris, J.S., Frith, C.D., Perrett, D.I., & Dolan, R. (1999). Dissociable neural responses to facial expressions of sadness and anger. *Brain*, **122**, 883-893.

Blair, R.J.R., Sellars, C., Strickland, I., Clark, F., Williams, A.O., Smith, M., & Jones, L. (1995). Emotion attributions in the psychopath. *Personality and Individual Differences*, **19**, 431-437.

Blanchard, R.J., Blanchard, D.C., & Takahashi, L.K. (1977). Attack and defensive behavior in the albino rat. *Animal Behavior*, **25**, 197-224.

Blumer, D., & Benson, D.F. (1975). Personality changes with frontal and temporal lobe lesions. In D.F. Benson, & D. Blumer (Eds.), *Psychiatric aspects of neurological disease* (pp. 151-170). New York: Grune & Stratton.

Bond, A.J., Curran, H.V., Bruce, M., O'Sullivan, G., & Shine, P. (1995). Behavioral aggression in panic disorder after 8 weeks of treatment with alprazolam. *Journal of Affective Disorders*, **35**, 117-123.

Borrill, J.A., Rosen, B.K., & Summerfield, A.B. (1987). The influence of alcohol on judgment of facial expressions of emotion. *British Journal of Medical Psychology*, **60**, 71-77.

Burgess, P.W., & Wood, R.L. (1990). Neuropsychology of behavior disorders following brain injury. In R.L. Wood (Ed.), *Neurobehavioural sequelae of traumatic brain injury* (pp. 110-133). London: Taylor & Francis.

Camras, L.A. (1977). Facial expressions used by children in a conflict situation. *Child Development*, **48**, 1431-1435.

Chaplin, T.C., Rice, M.E., & Harris, G.T. (1995). Salient victim suffering and the sexual responses of child molesters. *Journal of Consulting and Clinical Psychology*, **63**, 249-255.

Christiansen, P., Toft, P., Larsson, H.B., Stubgaard, M., & Henriksen, O. (1993). The concentration of N-acetyl aspartate, creatine + phosphocreatine, and choline in different parts of the brain in adulthood and senium. *Magnetic Resonance Imaging*, **11**, 799-806.

Cleckley, H. (1967). *The mask of sanity* (4th ed.). St Louis, MO: Mosby.

Cornell, D.G., Warren, J., Hawk, G., Stafford, E., Oram, G., & Pine, D. (1996). Psychopathy in instrumental and reactive violent offenders. *Journal of Consulting and Clinical Psychology*, **64**, 783-790.

Crick, N.R., & Dodge, K.A. (1996). Social information-processing mechanisms on reactive and proactive aggression. *Child Development*, **67**, 993-1002.

Critchley, H.D., Simmons, A., Daly, E.M., Russell, A., van Amelsvoort, T., Robertson, D.M., Glover, A., & Murphy, D.G. (2000). Prefrontal and medial temporal correlates of repetitive violence to self and others. *Biological Psychiatry*, **47**, 928-934.

Damasio, A.R. (1994). *Descartes' error: Emotion, rationality and the human brain.* New York: Putnam (Grosset Books).

Damasio, A.R., Tranel, D., & Damasio, H. (1990). Individuals with sociopathic behavior caused by frontal damage fail to respond autonomically to social stimuli. *Behavioral Brain Research*, **41**, 81-94.

Damasio, A.R., Tranel, D., & Damasio, H.C. (1991). Somatic markers and the guidance of behavior: Theory and preliminary testing. In H.S. Levin, H.M. Eisenberg, & A.L. Benton (Eds.), *Frontal lobe function and dysfunction* (pp. 217-229). New York: Oxford University Press.

Davis, M. (2000). The role of the amygdala in conditioned and unconditioned fear and anxiety. In J.P. Aggleton (Ed.), *The amygdala: A functional analysis* (pp. 289-310). Oxford: Oxford University Press.

de Waal, F.B. (1992). Coalitions as part of reciprocal relations in the Arnhem chimpanzee colony. In A.H. Harcourt, & F.B.M. de Waal (Eds.), *Coalitions and alliances in humans and other animals* (pp. 233-257). Oxford: Oxford University Press.

Dias, R., Robbins, T.W., & Roberts, A.C. (1996). Dissociation in prefrontal cortex of affective and attentional shifts. *Nature*, **380**, 69-72.

Dougherty, D.D., Shin, L.M., Alpert, N.M., Pitman, R.K., Orr, S.P., Lasko, M., Macklin, M.L., Fischman, A.J., & Rauch, S.L. (1999). Anger in healthy men: a PET study using script-driven imagery. *Biological Psychiatry*, **46**, 466-472.

Eibl-Eibesfeldt, I. (1970). *Ethology: The biology of behavior.* New York: Holt, Rinehart and Winston.

Ekman, P., & Friesen, W.V. (1976). *Pictures of facial affect.* Palo Alto: Consulting Psychologists Press.

Everitt, B.J., Cardinal, R.N., Hall, J., Parkinson, J.A., & Robbins, T.W. (2000). Differential involvement of amygdala subsystems in appetitive conditioning and drug addiction. In J.P. Aggleton (Ed.), *The Amygdala: A functional analysis* (pp. 289-310). Oxford: Oxford University Press.

Fine, C. (2000). *Expectation violations and emotional learning.* Unpublished doctoral dissertation, University College London, University of London, London.

Fine, C., & Blair, R.J.R. (2000). Mini review: The cognitive and emotional effects of amygdala damage. *Neurocase*, **6**, 435-450.

Fisher, L., & Blair, R.J.R. (1998). Cognitive impairment and its relationship to psychopathic tendencies in children with emotional and behavioral difficulties. *Journal of Abnormal Child Psychology*, **26**, 511-519.

Fowles, D.C. (1988). Psychophysiology and psychopathy: A motivational approach. *Psychophysiology*, **25**, 373-391.

Frick, P.J., & Hare, R.D. (in press). *The psychopathy screening device*. Toronto: Multi-Health Systems.

Gallagher, M., McMahan, R.W., & Schoenbaum, G. (1999). Orbitofrontal cortex and representation of incentive value in associative learning. *Journal of Neuroscience*, **19**, 6610-6614.

Goyer, P.F., Andreason, P.J., Semple, W.E., Clayton, A.H., King, A.C., Compton-Toth, B.A., Schulz, S.C., & Cohen, R.M. (1994). Positron-emission tomography and personality disorders. *Neuropsychopharmacology*, **10**, 21-28.

Grafman, J., Schwab, K., Warden, D., Pridgen, B.S., & Brown, H.R. (1996). Frontal lobe injuries, violence, and aggression: A report of the Vietnam head injury study. *Neurology*, **46**, 1231-1238.

Graves, R., Landis, T., & Goodglass, H. (1981). Laterality and sex differences for visual recognition of emotional and non-emotional words. *Neuropsychologia*, **19**, 95-102.

Hare, R.D. (1965). Temporal gradient of fear arousal in psychopaths. *Journal of Abnormal Psychology*, **70**, 442-445.

Hare, R.D. (1978). Electrodermal and cardiovascular correlates of psychopathy. In R.D. Hare, & D. Schalling (Eds.), *Psychopathic behavior: Approaches to research* (pp. 107-143). Chichester, England: Wiley.

Hare, R.D. (1982). Psychopathy and physiological activity during anticipation of an aversive stimulus in a distraction paradigm. *Psychophysiology*, **19**, 266-271.

Hare, R.D. (1991). *The Hare psychopathy checklist-revised.* Toronto, Ontario: Multi-Health Systems.

Hare, R.D., Frazelle, J., & Cox, D.N. (1978). Psychopathy and physiological responses to threat of an aversive stimulus. *Psychophysiology*, **15**, 165-172.

Hare, R.D., & Jutai, J.W. (1983). Criminal history of the male psychopath: Some preliminary data. In K.T. Van Dusen, & S.A. Mednick (Eds.), *Prospective studies of crime and delinquency* (pp. 225-236). Boston: Kluwer-Nijhoff.

Harpur, T.J., & Hare, R.D. (1994). Assessment of psychopathy as a function of age. *Journal of Abnormal Psychology*, **103**, 604-609.

Hart, S.D., & Hare, R.D. (1996). Psychopathy and antisocial personality disorder. *Current Opinion in Psychiatry*, **9**, 129-132.

Hatfield, T., Han, J.S., Conley, M., Gallagher, M., & Holland, P. (1996). Neurotoxic lesions of basolateral, but not central, amygdala interfere with Pavlovian second-order conditioning and reinforcer devaluation effects. *Journal of Neuroscience*, **16**, 5256-5265.

Hecaen, H., & Albert, M.L. (1978). *Human neuropsychology*. New York: Wiley.

Hornak, J., Rolls, E.T., & Wade, D. (1996). Face and voice expression identification in patients with emotional and behavioral changes following ventral frontal damage. *Neuropsychologia*, **34**, 247-261.

House, T.H., & Milligan, W.L. (1976). Autonomic responses to modeled distress in prison psychopaths. *Journal of Personality and Social Psychology*, **34**, 556-560.

Intrator, J., Hare, R.D., Stritzke, P., Brichtswein, K., Dorfman, D., Harpur, T.J., Bernstein, D., Handelsman, L., Scaefer, C., Keilp, J., Rosen, J., & Machac, J. (1997). A brain imaging (single photon emission computerized tomography) study of semantic and affective processing in psychopaths. *Biological Psychiatry*, **42**, 96-103.

Kesler/West, M.L., Andersen, A.H., Smith, C.D., Avison, M.J., Davis, C.E., Kryscio, R.J., & Blonder, L.X. (2001). Neural substrates of facial emotion processing using fMRI. *Cognitive Brain Research*, **11**, 213-226.

Killcross, S., Robbins, T.W., & Everitt, B.J. (1997). Different types of fear-conditioned behavior mediated by separate nuclei within amygdala. *Nature*, **388**, 377-380.

Laakso, M.P., Vaurio, O., Koivisto, E., Savolainen, L., Eronen, M., Aronen, H.J., Hakola, P., Repo, E., Soininen, H., & Tiihonen, J. (2001). Psychopathy and the posterior hippocampus. *Behavior Brain Research*, **118**, 187-193.

LaBar, K.S., LeDoux, J.E., Spencer, D.D., & Phelps, E.A. (1995). Impaired fear conditioning following unilateral temporal lobectomy in humans. *Journal of Neuroscience*, **15**, 6846-6855.

LaPierre, D., Braun, C.M.J., & Hodgins, S. (1995). Ventral frontal deficits in psychopathy: Neuropsychological test findings. *Neuropsychologia*, **33**, 139-151.

LeDoux, J. (2000). The amygdala and emotion: a view through fear. In J.P. Aggleton (Ed.), *The amygdala: A functional analysis* (pp. 289-310). Oxford: Oxford University Press.

Levenston, G.K., Patrick, C.J., Bradley, M.M., & Lang, P.J. (2000). The psychopath as observer: Emotion and attention in picture processing. *Journal of Abnormal Psychology*, **109**, 373-386.

Linnoila, M., Virkkunen, M., Scheinin, M., Nuutila, A., Rimon, R., & Goodwin, F.K. (1983). Low cerebrospinal fluid 5-hydroxy indoleacetic acid concentration differentiates impulsive from nonimpulsive violent behavior. *Life Sciences*, **33**, 2609-2614.

Lorenz, K. (1966). *On aggression*. New York: Harcourt Brace Jovanovich.

Lykken, D.T. (1957). A study of anxiety in the sociopathic personality. *Journal of Abnormal and Social Psychology*, **55**, 6-10.

Lykken, D.T. (1995). *The antisocial personalities*. Hillsdale, New Jersey: Erlbaum.

Mealey, L. (1995). The sociobiology of sociopathy: An integrated evolutionary model. *Behavioral and Brain Sciences*, **18**, 523-599.

Mitchell, D.G.V., Colledge, E., Leonard, A., & Blair, R.J.R. (submitted). Somatic markers and response reversal: Is there evidence of orbitofrontal cortex dysfunction in psychopathic individuals?

Morris, J.S., Frith, C.D., Perrett, D.I., Rowland, D., Young, A.W., Calder, A.J., & Dolan, R.J. (1996). A differential response in the human amygdala to fearful and happy facial expressions. *Nature*, **383**, 812-815.

Morris, J.S., Scott, S.K., & Dolan, R.J. (1999). Saying it with feeling: neural responses to emotional vocalizations. *Neuropsychologia*, **37**, 1155-1163.

Murray, E.A., Bussey, T.J., & Wise, S.P. (2000). Role of prefrontal cortex in a network for arbitrary visuomotor mapping. *Experimental Brain Research*, **133**, 114-129.

Newman, J.P., & Kosson, D.S. (1986). Passive avoidance learning in psychopathic and nonpsychopathic offenders. *Journal of Abnormal Psychology*, **95**, 252-256.

Newman, J.P., Patterson, C.M., & Kosson, D.S. (1987). Response perseveration in psychopaths. *Journal of Abnormal Psychology*, **96**, 145-148.

Newman, J.P., & Schmitt, W.A. (1998). Passive avoidance in psychopathic offenders: a replication and extension. *Journal of Abnormal Psychology*, **107**, 527-532.

Nucci, L.P., & Herman, S. (1982). Behavioral disordered children's conceptions of moral, conventional, and personal issues. *Journal of Abnormal Child Psychology*, **10**, 411-425.

O'Brien, B.S., & Frick, P.J. (1996). Reward dominance: Associations with anxiety, conduct problems, and psychopathy in children. *Journal of Abnormal Child Psychology*, **24**, 223-240.

Ogloff, J.R., & Wong, S. (1990). Electrodermal and cardiovascular evidence of a coping response in psychopaths. *Criminal Justice and Behavior*, **17**, 231-245.

Panksepp, J. (1998). *Affective neuroscience: The foundations of human and animal emotions*. New York: Oxford University Press.

Passingham, R.E., & Toni, I. (2001). Contrasting the dorsal and ventral visual systems: guidance of movement versus decision making. *Neuroimage*, **14**, S125-S131.

Passingham, R.E., Toni, I., & Rushworth, M.F. (2000). Specialisation within the prefrontal cortex: the ventral prefrontal cortex and associative learning. *Experimental Brain Research*, **133**, 103-113.

Patrick, C.J. (1994). Emotion and psychopathy: Startling new insights. *Psychophysiology*, **31**, 319-330.

Patrick, C.J., Bradley, M.M., & Lang, P.J. (1993). Emotion in the criminal psychopath: Startle reflex modulation. *Journal of Abnormal Psychology*, **102**, 82-92.

Patrick, C.J., Cuthbert, B.N., & Lang, P.J. (1994). Emotion in the criminal psychopath: Fear image processing. *Journal of Abnormal Psychology*, **103**, 523-534.

Pennington, B.F., & Bennetto, L. (1993). Main effects or transaction in the neuropsychology of conduct disorder? Commentary on "The neuropsychology of conduct disorder". *Development and Psychopathology*, **5**, 153-164.

Perry, D.G., & Perry, L.C. (1974). Denial of suffering in the victim as a stimulus to violence in aggressive boys. *Child Development*, **45**, 55-62.

Phelps, E.A., O'Connor, K.J., Gatenby, J.C., Gore, J.C., Grillon, C., & Davis, M. (2001). Activation of the left amygdala to a cognitive representation of fear. *Nature Neuroscience*, **4**, 437-441.

Phillips, M.L., Young, A.W., Scott, S.K., Calder, A.J., Andrew, C., Giampietro, V., Williams, S.C., Bullmore, E.T., Brammer, M., & Gray, J.A. (1998). Neural responses to facial and vocal expressions of fear and disgust. *Proceedings of the Royal Society of London.B*, **265**, 1809-1817.

Phillips, M.L., Young, A.W., Senior, C., Brammer, M., Andrews, C., Calder, A.J., Bullmore, E.T., Perrett, D.I., Rowland, D., Williams, S.C.R., Gray, J.A., & David, A.S. (1997). A specified neural substrate for perceiving facial expressions of disgust. *Nature*, **389**, 495-498.

Rahman, S.J., Sahakian, B.N., Cardinal, R., Rogers, R., & Robbins, T. (2001). Decision making and neuropsychiatry. *Trends in Cognitive Science*, **5**, 271-277.

Rahman, S., Sahakian, B.J., Hodges, J.R., Rogers, R.D., & Robbins, T.W. (1999). Specific cognitive deficits in mild frontal variant frontotemporal dementia. *Brain*, **122**, 1469-1493.

Raine, A., Buchsbaum, M.S., & LaCasse, L. (1997). Brain abnormalities in murderers indicated by positron emission tomography. *Biological Psychiatry*, **42**, 495-508.

Raine, A., Buchsbaum, M.S., Stanley, J., Lottenberg, S., Abel, L., & Stoddard, J. (1994). Selective reductions in prefrontal glucose metabolism in murderers. *Biological Psychiatry*, **36**, 365-373.

Raine, A., Lencz, T., Bihrle, S., LaCasse, L., & Colletti, P. (2000). Reduced prefrontal gray matter volume and reduced autonomic activity in antisocial personality disorder. *Archives of General Psychiatry*, **57**, 119-127.

Raine, A., Phil, D., Stoddard, J., Bihrle, S., & Buchsbaum, M. (1998). Prefrontal glucose deficits in murderers lacking psychosocial deprivation. *Neuropsychiatry, Neuropsychology and Behavioral Neurology*, **11**, 1-7.

Rapcsak, S.Z., Galper, S.R., Comer, J.F., Reminger, S.L., Nielsen, L., Kaszniak, A.W., Verfaellie, M., Laguna, J.F., Labiner, D.M., & Cohen, R.A. (2000). Fear recognition deficits after focal brain damage: A cautionary note. *Neurology*, **54**, 575-581.

Rogers, R.D., Everitt, B.J., Baldacchino, A., Blackshaw, A.J., Swainson, R., Wynne, K., Baker, N.B., Hunter, J., Carthy, T., Booker, E., London, M., Deakin, J.F., Sahakian, BJ., & Robbins, T.W. (1999). Dissociable deficits in the decision-making cognition of chronic amphetamine abusers, opiate abusers, patients with focal damage to prefrontal cortex, and tryptophan-depleted normal volunteers: evidence for monoaminergic mechanisms. *Neuropsychopharmacology*, **20**, 322-339.

Rogers, R.D., Owen, A.M., Middleton, H.C., Williams, E.J., Pickard, J.D., Sahakian, B.J., & Robbins, T.W. (1999). Choosing between small, likely rewards and large, unlikely rewards activates inferior and orbital prefrontal cortex. *Journal of Neuroscience*, **19**, 9029-9038.

Rogers, R.D., & Robbins, T.W. (2001). Investigating the neurocognitive deficits associated with chronic drug misuse. *Current Opinions in Neurobiology*, **11**, 250-257.

Rolls, E.T. (1997). The orbitofrontal cortex. *Philosophical Transactions of the Royal Society, B*, **351**, 1433-1443.

Rolls, E.T., & Stringer, S.M. (2001). A model of the interaction between mood and memory. *Network*, **12**, 89-109.

Schalling, D. (1978). Psychopathy-related personality variables and the psychophysiology of socialization. In R.D. Hare, & D. Schalling (Eds.), *Psychopathic behavior: Approaches to research* (pp. 85-106). Chichester, England: Wiley.

Schneider, F., Gur, R.E., Mozley, L.H., Smith, R.J., Mozley, P.D., Censits, D.M., Alavi, A., & Gur, R.C. (1995). Mood effects on limbic blood flow correlate with emotional self-rating: a PET study of oxygen-15 labeled water. *Psychiatry Research: Neuroimaging*, **61**, 265-283.

Schneider, F., Habel, U., Kessler, C., Posse, S., Grodd, W., & Muller-Gartner, H.W. (2000). Functional imaging of conditioned aversive emotional responses in antisocial personality disorder. *Neuropsychobiology*, **42**, 192-201.

Schneider, F., Weiss, U., Kessler, C., Muller-Gartner, H.W., Posse, S., Salloum, J.B., Grodd, W., Himmelmann, F., Gaebel, W., & Birbaumer, N. (1999). Subcortical correlates of differential classical conditioning of aversive emotional reactions in social phobia. *Biological Psychiatry*, **45**, 863-871.

Schoenbaum, G., Chiba, A.A., & Gallagher, M. (1998). Orbitofrontal cortex and basolateral amygdala encode expected outcomes during learning. *Nature Neuroscience*, **1**, 155-159.

Schoenbaum, G., Chiba, A.A., & Gallagher, M. (2000). Changes in functional connectivity in orbitofrontal cortex and basolateral amygdala during learning and reversal training. *Journal of Neuroscience*, **20**, 5179-5189.

Scott, S.K., Young, A.W., Calder, A.J., Hellawell, D.H., Aggleton, J.P., & Johnson, M. (1997). Impaired auditory recognition of fear and anger following bilateral amygdala lesions. *Nature*, **385**, 254-257.

Shapiro, S.K., Quay, H.C., Hogan, A.E., & Schwartz, K.P. (1988). Response perseveration and delayed responding in undersocialised aggressive conduct disorder. *Journal of Abnormal Psychology*, **97**, 371-373.

Siegel, A., Roeling, T.A., Gregg, T.R., & Kruk, M.R. (1999). Neuropharmacology of brain-stimulation-evoked aggression. *Neuroscience Biobehavioral Review*, **23**, 359-389.

Smetana, J.G. (1993). Understanding of social rules. In M. Bennett (Ed.), *The child as psychologist: An introduction to the development of social cognition* (pp. 111-141). New York: Harvester Wheatsheaf.

Sprengelmeyer, R., Rausch, M., Eysel, U.T., & Przuntek, H. (1998). Neural structures associated with the recognition of facial basic emotions. *Proceedings of the Royal Society of London.B*, **265**, 1927-1931.

Stevens, D., Charman, T., & Blair, R.J.R. (2001). Recognition of emotion in facial expressions and vocal tones in children with psychopathic tendencies. *Journal of Genetic Psychology*, **162**, 201-211.

Stone, V.E., Baron-Cohen, S., & Knight, R.T. (1998). Frontal lobe contributions to theory of mind. *Journal of Cognitive Neuroscience*, **10**, 640-656.

Strauss, E. (1983). Perception of emotional words. *Neuropsychologia*, **21**, 99-103.

Stuss, D.T., & Benson, D.F. (1986). *The frontal lobes*. New York: Raven Press.

Thorpe, S.J., Rolls, E.T., & Maddison, S. (1983). The orbitofrontal cortex: neuronal activity in the behaving monkey. *Experimental Brain Research*, **49**, 93-115.

Tranel, D., & Damasio, H. (1994). Neuroanatomical correlates of electrodermal skin conductance responses. *Psychophysiology*, **31**, 427-438.

Urenjak, J., Williams, S.R., Gadian, D.G., & Noble, M. (1993). Proton nuclear magnetic resonance spectroscopy unambiguously identifies different neural cell types. *Journal of Neuroscience*, **13**, 981-989.

Volavka, J. (1995). *Neurobiology of violence*. Washington, DC: American Psychiatric Press.

Volkow, N.D., & Tancredi, L. (1987). Neural substrates of violent behavior. A preliminary study with positron emission tomography. *British Journal of Psychiatry*, **151**, 668-673.

Volkow, N.D., Tancredi, L.R., Grant, C., Gillespie, H., Valentine, A., Mullani, N., Wang, G.J., & Hollister, L. (1995). Brain glucose metabolism in violent psychiatric patients: a preliminary study. *Psychiatry Research*, **61**, 243-253.

Whalen, P.J., Shin, L.M., McInerney, S.C., & Rauch, S.L. (1998). Greater fMRI activation to fearful vs. angry expressions in the amygdaloid region. *Neuroscience Abstracts*, **24**, 692.

Williamson, S., Hare, R.D., & Wong, S. (1987). Violence: Criminal psychopaths and their victims. *Canadian Journal of Behavioral Science*, **19**, 454-462.

Williamson, S., Harpur, T.J., & Hare, R.D. (1991). Abnormal processing of affective words by psychopaths. *Psychophysiology*, **28**, 260-273.

Wong, M., Fenwick, P., Fenton, G., Lumsden, J., Maisey, M., & Stevens, J. (1997). Repetitive and non-repetitive violent offending behavior in male patients in a maximum security mental hospital—clinical and neuroimaging findings. *Medicine, Science and Law*, **37**, 150-160.

Chapter 8

PSYCHOPHYSIOLOGICAL CORRELATES OF ANTISOCIAL BEHAVIOR: A CENTRAL CONTROL HYPOTHESIS

Sharon S. Ishikawa, Ph.D. and Adrian Raine, D.Phil.
Department of Psychology, University of Southern California

Acknowledgments. This research was conducted with the support of a National Research Service Award (1 F32 MH12951-01) to the first author and an Independent Scientist Award (K02 MH01114) to the second author.

INTRODUCTION

Psychophysiological processes—or the study of cognitive, emotional, and behavioral phenomena in relation to physiological principles (Cacioppo & Tassinary, 1990)—represent one of the most broadly studied biological correlates of antisocial behavior. Traditionally, measures of central (i.e., electroencephalogram, event-related potentials) and autonomic (i.e., heart rate, skin conductance) nervous system functions are discussed as separate entities. However extensive research—primarily with animals, but also with humans — has helped delineate the central nervous system control of autonomic functions. Because several reviews on the psychophysiology of antisocial behavior have been previously published (e.g., Fowles, 1993; Lahey, McBurnett, Loeber, & Hart, 1995), the purpose of the current chapter is to update the prior reviews and integrate the findings with emerging developments from the fields of behavioral and affective neuroscience. Although the chapter places the findings within the context of a central control hypothesis, it should be stated from the outset that it is not assumed that the hypothesized neuroanatomical correlates from one area (e.g., electroencephalogram) directly correspond to the neuroanatomical components of another (e.g., skin conductance).

CENTRAL NERVOUS SYSTEM

Overview

The central nervous system is composed of the brain and spinal cord and regulates and integrates information from the peripheral nervous system. The cortex or surface area of the brain is divided lengthwise into two hemispheres (i.e., right, left), which are further subdivided bilaterally into four different lobes (i.e., frontal, temporal, parietal, occipital; see Figure 1). Each lobe is generally associated with specific functions. Due to its mediation of cognitive, affective, and motivational processes, the prefrontal area of the frontal lobe is of particular interest to the study of antisocial behavior (Cummings, 1995). The prefrontal area consists of the dorsolateral, orbital, and medial (or anterior cingulate) cortices. Each of these prefrontal cortices (PFCs) is part of a frontal-subcortical circuit that integrates information from throughout the brain to organize behavior (Cummings, 1995). Although each of the prefrontal circuits is complex, only a brief description of some basic elements common to antisocial behavior and psychophysiological functioning are addressed below.

The *dorsolateral* prefrontal cortex (see Figure 1) is extensively interconnected with the cortical association areas and the orbital prefrontal cortex. It is critical to information processing, especially as it relates to working memory (i.e., the ability to hold multiple representations in mind for use in problem-solving) and executive functions (i.e., a cluster of higher-order cognitive processes involving initiation, planning, cognitive flexibility, abstraction, and decision-making; Cummings, 1995; Damasio & Anderson, 1993). The *orbital* prefrontal cortex (see Figure 2) is extensively interconnected with the

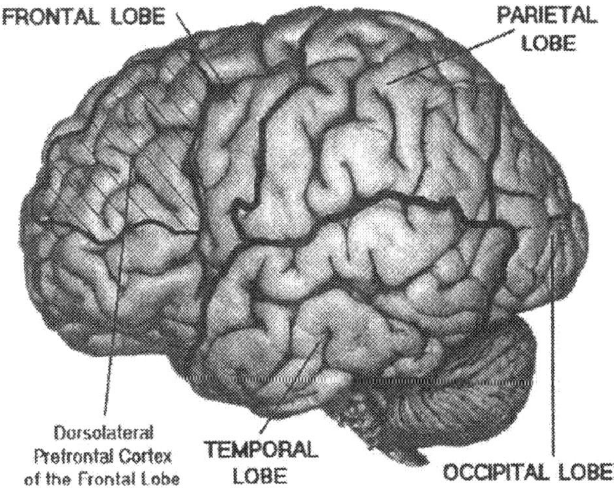

Figure 1. Delineation of the human cortical lobes (left hemisphere). Dorsolateral prefrontal cortex within the frontal lobe is indicated by hashed lines. Adapted from Davidson, Putnam, & Larson (2000), and DeArmond, Fusco, & Dewey (1989). Copyright (2000) American Association for the Advancement of Science, and used by permission of Oxford University Press, Inc.

amygdala and temporal pole, and regulates behavioral inhibition, autonomic reactivity, social and self-awareness, and aggression (Damasio & Anderson, 1993; LaPierre, Braun, & Hodgins, 1995). It also contains part of the ventromedial prefrontal cortex, which is implicated in risk-related and emotion-based decision-making (Bechara, Damasio, Damasio, & Lee, 1999). The *medial* prefrontal cortex (see Figure 3) is interconnected with the dorsolateral prefrontal cortex and contains the remaining portion of the ventromedial prefrontal cortex. The medial PFC projects, in part, from the anterior cingulate through the nucleus accumbens and ventral pallidum to the hypothalamus and amygdala, and mediates motivational processes and maintenance of activity (Cummings, 1995).

Also contained within each hemisphere are the temporal, parietal, and occipital lobes (see Figure 1). The *temporal lobe* is largely involved in hearing, understanding the content and emotional intonation of spoken language, and the storing of visual memories. The medial temporal lobe contains the hippocampus, which is involved in memory functions, and the amygdala, which is involved in emotion and the formation of associations between environmental stimuli and emotional responses. The *parietal lobe* contains the somatosensory cortex, which makes it central to processing bodily sensations. The parietal lobe also integrates sensory information across modalities (i.e., auditory, visual, etc.), and is involved in visual attention. Finally, the occipital lobe is divided into different areas responsible for visual function.

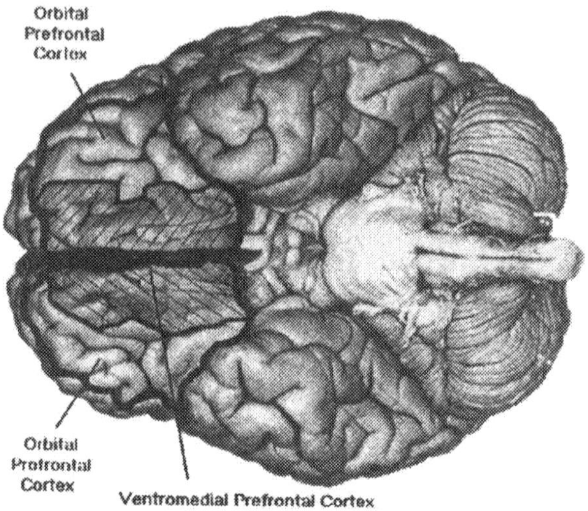

Figure 2. *View of the underside of the brain with the orbital prefrontal cortex (OPC) delineated bilaterally. Contained within the OPC is part of the ventromedial prefrontal cortex (indicated by hashed lines). Adapted from Davidson, Putnam, & Larson (2000), and DeArmond, Fusco, & Dewey (1989). Copyright (2000) American Association for the Advancement of Science, and used by permission of Oxford University Press, Inc.*

Central Psychophysiological Measures: EEG and ERPs

With regard to psychophysiological measures of central function, the most commonly utilized methods are the electroencephalogram (EEG) and event related potentials (ERPs). More recently, non-invasive functional brain-imaging techniques (e.g., positron emission tomography or PET; fMRI or functional magnetic resonance imaging) have begun to be used. Findings from EEG and ERP research are reviewed below, and brain imaging studies are reviewed at the conclusion of the chapter.

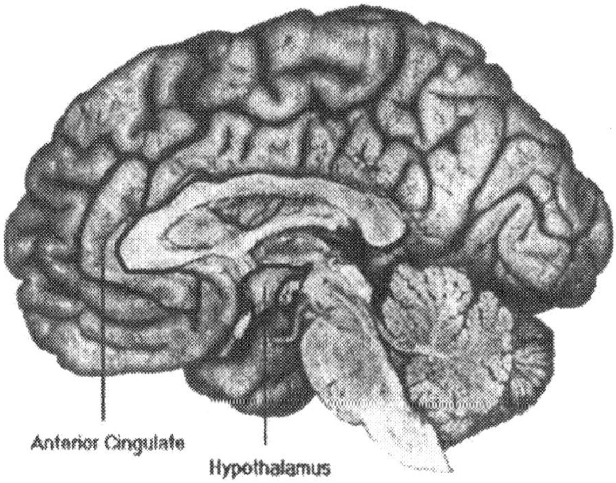

Figure 3. Left hemisphere view of the anterior cingulate, which is contained within the medial prefrontal cortex, and the hypothalamus. Adapted from Davidson, Putnam, & Larson (2000), and DeArmond, Fusco, & Dewey (1989). Copyright (2000) American Association for the Advancement of Science, and used by permission of Oxford University Press, Inc.

Electroencephalogram

Overview

The electroencephalogram (EEG) is a non-invasive recording of brain electrical activity via a standardized array of surface electrodes attached to an individual's scalp. The signal is recorded as continuous, sinusoidal oscillations across different power bands (e.g., alpha, theta) and is thought to result from postsynaptic potentials summated across large groups of neurons. The different power bands are associated with different degrees of consciousness or arousal (see Table 1). Awake individuals show activity in all bands, although there are individual differences in the power of each band (Yaralian & Raine, 2000). For example, individuals with high levels of cortical activity typically show a predominance of fast alpha and beta waves relative to individuals with low cortical arousal, who typically show a predominance of theta, delta, and slow alpha waves. Despite the theoretical importance of gamma activity, it is particularly susceptible to movement artifact and is considered unreliable unless adequate data pre-processing has been employed. To date it has not been reported in any studies on antisocial behavior, and it will thus not be included in the present review.

Table 1. Decomposition of EEG power bands.

Power Band	Lower –Upper Boundaries	General Arousal Level
Delta	0 - 4 Hz	sleep
Theta	5 - 7 Hz	low arousal and alertness
Alpha	8 - 12 Hz	relaxed wakefulness
(slow alpha)	(7.5 - 9.5 Hz)	(low arousal)
(fast alpha)	(9.5 - 12.5 Hz)	(high arousal)
Beta	13 - 20 or 13 - 30 Hz	alertness and vigilance
Gamma	36 - 44 Hz	focused attention, perceptual binding

EEG research offers several strengths and limitations. Relative to other functional measures of brain activity (e.g., PET, fMRI), EEG provides outstanding temporal resolution. However, the skull and neural tissue act as low-pass filters that diffuse electrical activity across a relatively large area of the scalp. Although the EEG can be used to infer the general location of neural activity within the brain (e.g., frontal, parietal), its spatial resolution is in the order of 10 millimeters. Thus, the activity detected at a particular region of the scalp cannot be linked to a discrete, underlying neuroanatomical location. In addition, much of the early EEG research on antisocial and criminal populations is based upon visually scored tracings rather than quantitative computerized scored data (Volavka, 1987). Approximately 10-15% of visually scored EEGs from normal individuals are classified as "abnormal," and these base rates should be kept in mind when studies report group differences between antisocial and healthy control populations (Yaralian & Raine, 2000). Despite these limitations, the EEG is extensively used as a measure of central nervous system arousal.

Empirical Findings

A large number of studies have attempted to link EEG tracings to criminal, violent, and/or psychopathic behavior. Despite considerable variability in sample characteristics and EEG recording and scoring techniques, there are fairly consistent findings that criminal groups have higher rates of EEG abnormalities than controls (Yaralian & Raine, 2000). There is also some suggestion that cortical underarousal is particularly over-represented among violent individuals (Hill & Pond, 1952; Mark & Ervin, 1970; Williams, 1969). For example, aggressive offenders have EEGs characterized by predominant theta activity (Williams, 1969), and frequency of violent acting out by psychiatric inpatients is associated with increased delta and decreased alpha (Convit, Czobor, & Volavka, 1991). Similarly, highly aggressive, male drug abusers exhibit increased theta and decreased alpha relative to low aggressive abusers (Fishbein *et al.*, 1989), and murderers exhibit increased delta and theta compared to normal controls (Gatzke-Kopp, Raine, Buchsbaum, & LaCasse, in press). Patients with intermittent explosive disorder or episodic dyscontrol are also disproportionately more likely to have abnormal slowing on the EEG compared to depressed and headache control patients (Drake, Hietter, & Pakalnis, 1992). Moreover, EEG arousal deficits in criminals are most frequently observed in the temporal (Drake *et al.*, 1992; Volavka, 1995; Wong, Lumsden, Fenton, & Fenwick, 1994) and frontal regions of the brain (Fishbein *et al.*, 1989; Mednick, Volavka, Gabrielli, & Itil, 1981).

Although the majority of studies are cross-sectional in nature, a few prospective, longitudinal studies demonstrate that EEG abnormalities precede the onset of significant criminal behavior. For example, cortical underarousal (i.e., greater activity in theta, delta, slow alpha) at age 15 is observed in those who are convicted for a crime by age 24 but not in those who avoid conviction (Raine, Venables, & Williams, 1990). Similarly, alpha slowing measured in young children and adolescents is related to repeat criminal offending measured several years later, but not to one-time criminal offending or non-offending (Mednick *et al.*, 1981; Volavka, 1987). These findings apply both to a high-risk sample (i.e., parents with schizophrenia or personality disorders; Volavka, 1987) and to a community-based, normal sample (Mednick *et al.*, 1981). Because alpha frequency increases between the ages of 5 and 20 as a part of normal cortical development, it is thought to reflect brain maturation. As such, evidence of alpha slowing in these studies is interpreted as reflecting a cortical developmental lag in those prone to recidivistic crime (Volavka, 1987).

Recent studies have shifted away from a model of cortical underarousal to test a more specific hypothesis adopted from the emotion regulation literature. This hypothesis assesses whether antisocial and normal individuals differ in arousal *across* hemispheres. For example, atypical frontal activation on EEG (right > left) is associated with externalizing or internalizing behavior problems when the child has a sociable or shy temperament, respectively (Fox, Schmidt, Calkins, Rubin, & Coplan, 1996). There also appear to be gender differences in what constitutes "abnormal" cortical activation. Consistent with a large body of research on negative emotional reactivity in females, healthy girls exhibit greater left than right frontal activation (i.e., slow alpha) at age 8 (and no asymmetry at age 4), whereas girls diagnosed with Oppositional Defiant Disorder (ODD) show greater right than left activation at both ages (Baving, Laucht, & Schmidt, 2000). In contrast, healthy boys exhibit greater right than left frontal activation at both ages, whereas ODD boys fail to show frontal asymmetry at either age. Whether this activation difference between boys remains stable across the lifespan has yet to be determined, however, as Antisocial Personality Disorder is associated with greater frontal fast alpha activity in the right—not the left—hemisphere in adult males (Deckel, Hesselbrock, & Bauer, 1996).

Summary

Overall, EEG abnormalities—particularly cortical underarousal and fronto-temporal abnormalities—are consistently observed in antisocial individuals. It also appears that the pattern of cortical underarousal is moderated by age and gender. Early findings were interpreted as reflecting developmental immaturity, the deleterious effect of brain injuries, or frontal disinhibition (Yaralian & Raine, 2000). However, these explanations are somewhat circular in that they do not address, for example, whether the individual is predisposed towards antisocial behavior as a result of the head injury, or whether the individual's antisocial lifestyle renders him more prone towards head injury (see Chapter 6 for a more detailed discussion of this issue). In contrast, an emerging theory from developmental studies suggests that greater right than left frontal EEG activation constitutes, in part, a measure of atypical emotional reactivity reflecting a general predisposition towards psychopathology (Baving *et al.*, 2000; Davidson, 1998). Whether this affective predisposition ultimately manifests itself in antisocial behavior is thought to depend on the influence of other social and biological factors, such as child temperament, parental criminal behavior and parental psychopathology. In addition, there is some suggestion that the abnormal activation pattern in young boys (i.e., no frontal asymmetry) is different from that observed in young girls and adult males and females.

Event-Related Potentials

Overview

In contrast to the continuous EEG, event-related potentials (ERPs) refer to the study of electrical brain potentials that are time-locked (to within milliseconds) of discrete events. ERPs reflect brain electrical activity in preparation for, or in response to, a discrete event. As such, they are interpreted as reflecting the psychophysiological correlates of specific psychological processes (Fabiani, Gratton, & Coles, 2000). ERPs are acquired using the same technology as EEG but are smaller in voltage, thus requiring a separate set of analytic and methodological techniques to extract the signal from background EEG activity. ERPs offer excellent temporal resolution of neural changes within the brain. As with EEG, ERPs have poor spatial resolution and can typically only be used to infer general location of changes in neuronal activity. However, for some components the neural generators (i.e., location within the brain responsible for the electrical changes detected on the scalp) have been identified (Hugdahl, 1995).

ERP components are classified according to the direction of the waveform, as well as the latency between stimulus presentation and time of ERP recording. A "P" and "N" refer to positive-going and negative-going waveforms, respectively, with positive-going forms traditionally depicted in a downward direction (see Figure 4). The numbers that follow the N or P refer to the average number of milliseconds relative to the stimulus presentation, or the order in which a waveform occurs. That is, "N100" (or N1) is a negative-going waveform that occurs approximately 100 ms after a stimulus (and is typically the first negative-going waveform), and "P300" (or P3) refers to a positive-going waveform that occurs approximately 300 ms after a stimulus.

ERP components can be broadly classified into one of two categories: *exogenous* and *endogenous* (see Table 2). Exogenous ERP components reflect sensory registration of the stimulus and are "obligatory" or automatically processed. That is, the magnitude of the response simply depends on the presence of the stimulus and does not require a higher or more complex level of cognitive processing (Hugdahl, 1995). Exogenous components include auditory brainstem

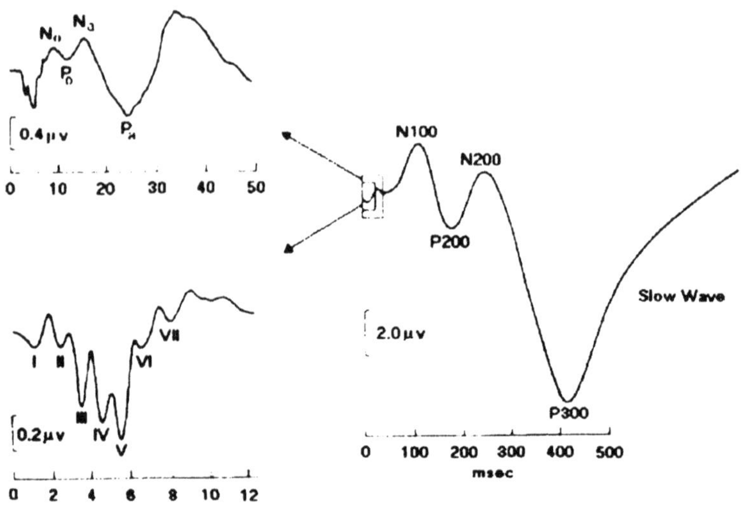

Figure 4. Representation of ERP components to auditory stimuli. The bottom left section represents auditory brainstem components; the top left section represents auditory middle latency components; and the right section represents long-latency exogenous and endogenous components. Reprinted from Fabiani, Gratton, & Coles (2000) with permission from the authors and Cambridge University Press.

potentials, middle-latency, and long-latency components. One of the most commonly studied long-latency components is N1, which is related to selective attention and provides a maximal response at the vertex (i.e., top center of the scalp). P2 is closely related to N1 but is not focally localized on the scalp (Hugdahl, 1995). One should note, however, that due to their association with selective attention, N1 and P2 are a blend of exogenous and endogenous components. Endogenous components also occur between 100-300 ms following stimulus presentation, but, in contrast to exogenous components, depend on the individual's level of higher cognitive processing (e.g., selective attention, responses to novel or unexpected stimuli). The most commonly studied endogenous ERP component is the P3. Although the P3 is traditionally measured at the vertex, it can also be measured across the frontal leads and is indicative of different neural generators depending on the recording site (see Table 2).

Empirical Findings

Relative to other psychophysiological measures, little ERP research using ERPs has been conducted with criminal populations. Within this limited literature base, early studies largely focused on a subset of criminals diagnosed as psychopaths, whereas more recent studies have focused on alcoholic antisocial individuals. Psychopaths represent a subset of antisocial individuals who—relative to individuals who commit crime or are diagnosed with Antisocial Personality Disorder—are particularly characterized by affective, interpersonal deficits such as glib superficiality, egocentricity, grandiosity, lack of remorse or guilt, lack of empathy, deceitful and manipulative behavior, and shallow affect (Hare, 1993). The early body of research on psychopaths has been extensively reviewed before (Raine, 1989). Key findings from the prior work are summarized below and individual studies not included in the original review are integrated as appropriate. One should note, however, that the conclusions drawn here are not necessarily the same as those from the earlier review (Raine, 1989).

The majority of early ERP studies on psychopaths examined long-latency exogenous and endogenous components (i.e., N1, positive slow wave or PSW). Results for long-latency exogenous components are mixed, with 6 of the 11 analyses from 9 different studies finding significant differences between psychopaths/antisocials and various controls (Raine, 1989). Due to the variability of scalp recording sites and ERP components used across studies (see Raine, 1989, for details), it is difficult to draw general conclusions. However, 4 of the 5 studies examining N1 in response to auditory stimuli were non-significant, suggesting that antisocials (psychopaths, in particular) do not differ from controls on cortical measures of selective auditory attention. In contrast, all 3 of the early studies examining N1 to visual stimuli

Table 2. ERP components and neural generators.

ERP Component	Timing (in ms.)	Function	Neural Generator
Exogenous			
(Auditory) Brainstem: I-VII	10-12	Sensory registration of sound	Cranial nerves, Brainstem, Thalamus (?)
(Auditory) Middle Latency: N0, P0, Na, Pa, Nb	>10-12	Sensory registration of sound	Heschl's gyrus, Thalamus
Long Latency:	100-300		
auditory N1, P2, (P1)	100, 200	Selective auditory attention	Primary auditory cortex in temporal lobe
visual P1	100	Sensory registration, selective visual attention	Occipital lobe
Endogenous			
Long Latency:	100-300		
P3 (vertex)	300	Controlled cognitive processing	Parietal and Temporal lobes
P3 (frontal)	300		Frontal lobe

showed significant effects (i.e., higher amplitude in psychopaths). It was initially hypothesized that this pattern of cortical augmenting reflects the psychophysiological correlate of stimulation- or sensation-seeking (see Chapters 2 and 3 for a discussion of sensation-seeking behavior). Although subsequent research has failed to support this hypothesis in psychopaths (Raine & Venables, 1990), initial findings suggest that non-psychopathic criminals, relative to normal controls, do appear to cortically augment visual stimuli (Raine, 1997; Raine & Venables, 1990), which is consistent with a sensation-seeking hypothesis.

The overall findings for early studies on long-latency endogenous components are similarly mixed, again making if difficult to draw general conclusions. A prior review suggested that if one examines the results based on the nature of the test paradigm, more coherent patterns emerge. Specifically, early work on vertex P3 responses to passive tasks such as orienting or aversive conditioning was characterized as inconsistent (i.e., non-significant, or significant but in opposite directions across studies; Raine, 1989). In contrast, early work on vertex, parietal, and frontal P3 or positive slow wave (PSW) responses during effortful information-processing tasks with a motor response (i.e., Go/No-Go, CPT) was characterized as more consistent (Raine, 1989), with psychopaths showing enhanced PSW and/or larger P3 amplitude compared to non-psychopaths (Jutai, Hare, & Connolly, 1987; Raine & Venables, 1988). As with the visual N1 studies reviewed above, these findings were originally interpreted as reflecting an enhanced ability of psychopaths to attend to things of interest (Raine, 1989). However, later work found that psychopaths and non-psychopaths do not differ on P300 amplitude to auditory phonemes while playing an interesting video game (although their PSWs were longer; Jutai, Hare, & Connolly, 1987), and that psychopaths show reduced P300 amplitude responses to a visual oddball task compared to non-psychopaths (Kiehl, Hare, Liddle, & McDonald, 1999). Thus, this hypothesis has not found consistent support.

An alternative hypothesis suggested by these early ERP studies is that the greater P3 amplitude or slow wave recovery in psychopaths simply reflects abnormal information processing. In a series of language-processing studies, Hare and colleagues observed that criminal psychopaths exhibit a large, atypical centrofrontal negative-going potential (Kiehl, Hare, McDonald, & Brink, 1999; Williamson, Harpur, & Hare, 1991). In addition, psychopaths fail to exhibit ERP differentiation of word types such as affective versus neutral words (Kiehl *et al.*, 1999; Williamson, Harpur, & Hare, 1991), or concrete versus abstract words (Kiehl *et al.*, 1999). Hare and colleagues interpret these findings to indicate impaired language processing in psychopaths (Kiehl *et al.*, 1999; Williamson, Harpur, & Hare, 1991). However, psychopaths also exhibit similar ERP responses to a non-language, visual oddball task (Kiehl, Hare, Liddle, & McDonald, 1999). In addition, on a non-language Go/No-Go task, psychopaths' P375 responses to the Go versus No-Go conditions are opposite of the non-psychopathic criminals, which are significantly smaller on Go than on No-Go (Kiehl, Smith, Hare, & Liddle, 2000). Thus, the hypothesized cognitive deficits in psychopaths may not necessarily be isolated to language.

More recent studies have moved away from studying psychopaths to studying alcoholics who are aggressive or have a comorbid diagnosis of Antisocial Personality Disorder (APD). In general, it appears that antisocial behavior problems are inversely associated with P300 amplitudes. Small frontal P300 amplitudes are related to impulsive aggression in normal college students (Gerstle, Mathias, & Stanford, 1998. Small P300s are also observed in prison inmates diagnosed only with APD (Barratt, Stanford, Kent, & Felthous, 1997), and in alcoholics with a history of incarceration for violent offenses (Branchley, Buydens-Branchley, & Lieber, 1988). In addition, adults with APD exhibit significantly smaller frontal P300 amplitudes to target stimuli compared to normal controls (Bauer, O'Connor, & Hesselbrock, 1994). This frontal P300-APD link is also observed in alcoholic adults, although the deficits are specific to antisocial individuals under age 30 (Costa *et al.*, 2000). Further evidence of age-related effects for P3 comes from Bauer and Hesselbrock (1999), who find that younger, conduct disordered (CD) adolescents exhibit the P300 deficit in the posterior region of the brain, whereas older CD adolescents exhibit the deficit in the anterior region.

Summary

Clearly, the overall findings from ERP research on antisocial behavior are disparate. However, relative consistency can be observed when the research is organized around measurement of a particular ERP component and/or pattern of ERP responding within a specific population (i.e., information processing in psychopaths; frontal P300 amplitudes in alcoholic APDs and impulsively aggressive individuals). Specifically, (a) psychopaths exhibit an atypical centrofrontal negative-going potential and fail to show ERP target/non-target differentiation during information-processing tasks, and (b) aggressive/

alcoholic/ antisocial individuals exhibit significantly smaller frontal P3 than non-antisocial controls. There is also some suggestion that frontal ERP deficits are most likely to be detected between late adolescence and young adulthood. While this coherence is an important first step, future research should focus on elucidating the functional and developmental significance of these findings. In addition, a number of methodological differences will need to be addressed to determine whether the discrepancy in ERP findings across these two antisocial populations is meaningful (e.g., psychopaths represent a unique subset of antisocial individuals) or artifactual (e.g., psychopathy studies have tended to look at P3 in nonfrontal sites, especially vertex, whereas alcohol-APD studies have primarily looked at frontal P3).

AUTONOMIC NERVOUS SYSTEM

Overview

The autonomic nervous system (ANS) serves as a link between the central nervous system (CNS) and the internal organs over which an individual ordinarily does not have control (e.g., heart, lungs, sweat glands, salivary glands, etc.). The ANS is regulated by complex interactions between the sympathetic nervous system (SNS), which mobilizes the body for action when threatened or under stress (i.e., the "fight-or-flight" response), and the parasympathetic nervous system (PNS), which controls vegetative functions and energy conservation (Spence & Mason, 1992). Although the sympathetic and parasympathetic systems are traditionally viewed as exerting reciprocal, antagonistic actions on each other, there is some evidence that the two systems are uncoupled (i.e., change in one does not affect the other) or coactive (i.e., both can be simultaneously active or inhibited) in their regulation of some organs (e.g., heart; Ohman, Hamm, & Hugdahl, 2000). With regard to the study of antisocial behavior, heart rate and skin conductance (a measure of electrodermal activity or sweating) are the most frequently used measures. Heart rate is controlled by both the PNS and SNS, whereas skin conductance is controlled by the SNS.

The Neuroanatomy of Autonomic Functions

Extensive research—particularly experimental research with animals—has helped delineate the complex network of brain and spinal cord structures responsible for regulating cardiovascular and electrodermal function. A complete discussion of these networks is beyond the scope of the current chapter. Instead, a brief outline of the structures fundamental to autonomic regulation is provided, with particular attention paid to the neuroanatomical structures that are also implicated in violent and antisocial behavior.

Hypothalamus

In essence, the hypothalamus is the brain's "command center" with regard to autonomic functioning (Sequeira & Roy, 1993), and it is centrally involved in organizing the behavioral and somatic responses that prepare an organism for *defensive* or *vigilance* reactions (see Figures 3, 5 for neuroanatomical location). The defense reaction is thought to facilitate active behavioral coping commonly known as the fight-or-flight response. Stimulation of specific regions of the hypothalamus elicits characteristic physiological responses such as increased heart rate, systemic arterial blood pressure, and sweating (Allen & Cecheto, 1992; Sequeira & Roy, 1993). The vigilance reaction, on the other hand, is associated with inhibitional behavioral coping such as freezing or orienting head motion, and is accompanied by cardiovascular changes such as a decrease in heart rate and blood pressure and facilitation of the baroreceptor reflex (i.e., modulation of heart rate in response to blood pressure changes and vasoconstriction). Although the hypothalamus serves a common integrative function for these various responses, differentiation of the cardiac vigilance, cardiac defense, and electrodermal defense reactions occurs in the descending pathways from the hypothalamus to the brainstem and spinal cord (Brownley, Hurwitz, & Schneiderman, 2000; Sequeira & Roy, 1993).

Temporal-Limbic Structures

The amygdala and hippocampus, located within the temporal lobe, also assist in the regulation of autonomic functions (see Figure 5). In animals, stimulation of the amygdala and temporal cortex consistently elicits the cardiac vigilance response (Brownley, Hurwitz, & Schneiderman, 2000). Similarly, the amygdala exerts an excitatory influence on electrodermal activity (Sequeira & Roy, 1993). In humans, size of the temporal area (including the amygdala) relates to skin conductance responses to novel stimuli (Raine, Reynolds, & Sheard, 1991). Damage to the amygdala and/or medio-basal temporal lobe frequently (Bechara, Damasio, Damasio, & Lee, 1999; Sequeira & Roy, 1993), but not always (Tranel & Damasio, 1989), disrupts the generation of skin conductance responses.

In contrast, electrical stimulation studies of the hippocampus in animals and humans have produced contradictory results on electrodermal activity, suggesting both an inhibitory and excitatory influence (Frederickson *et al.*, 1998). Moreover, it appears that the influence of the hippocampus on electrodermal activity is controlled by higher cortical structures, which are reviewed below.

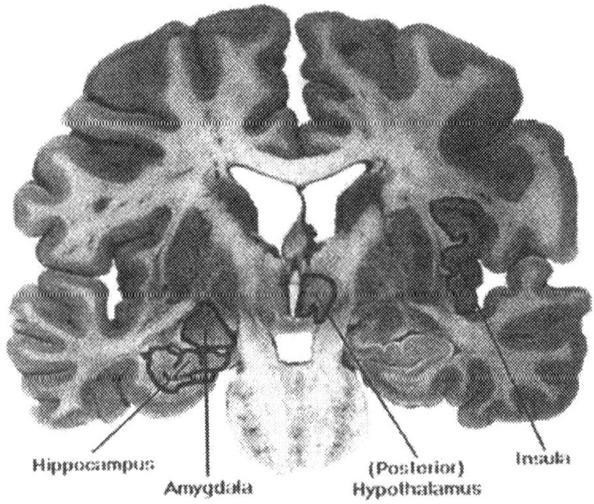

Figure 5. Coronal slice delineating the hippocampus, amygdala, posterior hypothalamus, and insula (each structure delineated in one hemisphere only). Adapted from DeArmond, Fusco, & Dewey (1989). Used by permission of Oxford University Press, Inc.

Cortical Structures

The influences of the human frontal cortex on autonomic activity are most relevant to psychological processes such as effortful information processing and classical conditioning (Frederickson *et al.*, 1998; Tranel & Damasio, 1994; Zahn, Grafman, & Tranel, 1999) rather than basic functions such as thermoregulation or unconditioned responses to aversive stimuli (Grueninger, Kimble, Grueninger, &

Levine, 1965). The key prefrontal structures that interact with the hypothalamus, hippocampus, and/or amygdala to regulate electrodermal and cardiovascular reactivity include the orbital PFC, dorsolateral PFC, and anterior cingulate (Bechara, Tranel, Damasio, & Damasio, 1996; Critchley, Elliott, Mathias, & Dolan, 2000; Grueninger *et al.*, 1965; Larsen, Schneiderman, & Pasin, 1986; Raine & Lencz, 1993; Raine *et al.*, 1991).

In addition, there is mounting evidence that the posterior insular cortex, which innervates the lateral hypothalamus (Allen & Cechetto, 1992), plays an important role in mediating cardiovascular function (Critchley, Corfield, Chandler, Mathias, & Dolan, 2000; Williamson *et al.*, 1997; Zhang, Dougherty, & Oppenheimer, 1999). It further appears that the left and right posterior insula are associated with regulation of parasympathetic and sympathetic cardiovascular function, respectively (Oppenheimer, Gelb, Girvin, & Hachinski, 1992; Zhang & Oppenheimer, 1998, 2000). See Figure 5 for neuroanatomical location of the insula.

One potential limitation of most of the above findings is that the results are derived from lesion or electrical stimulation studies with animals or with humans undergoing neurosurgery. To complement this research base, human functional imaging studies are beginning to identify the neuroanatomical structures involved in autonomic changes during complex cognitive tasks. For example, activation of the right orbital and dorsolateral PFCs (in conjunction with other frontal and medial temporal cortices) correspond to electrodermal responses during the extinction phase of aversive conditioning (Hugdahl, 1998). During exposure to mild effortful stressors, increases in heart rate correspond to significant activation changes in the bilateral orbital PFC, right cingulate, left insula and amygdala, as well as the pons, midline cerebellum, and right middle frontal gyrus (Critchley, Corfield, *et al.*, 2000). Thus, the emerging functional imaging work in normal humans is consistent with the neuropsychiatric and animal literatures.

Skin Conductance and Criminal Behavior

Overview

Measures of skin conductance (SC) activity capture small fluctuations in the electrical activity of the skin, with increased sweating causing increased SC activity. Although there are several possible skin conductance measures, only the ones reported in the literature on criminal behavior are reviewed here. At the broadest level, skin conductance can be measured when the person is at rest and not engaged in any particular task (i.e., resting or tonic skin conductance level, SCL) or when skin conductance fluctuates in response to a stimulus (i.e., skin conductance response, SCR). SCRs, in turn, can be broken down in a number of ways. The most commonly presented components in the literature reviewed below are SCR amplitude and SCR frequency. Amplitude refers to the increase in SC measured from baseline to response peak (see Figure 6), and frequency refers to the number of measurable SCRs following a stimulus (Hugdahl, 1995). In addition, the frequency of non-specific skin conductance responses (NSRs) is occasionally reported, which refers to SC responses that occur in the absence of an identified stimulus (Dawson, Schell, & Filion, 2000).

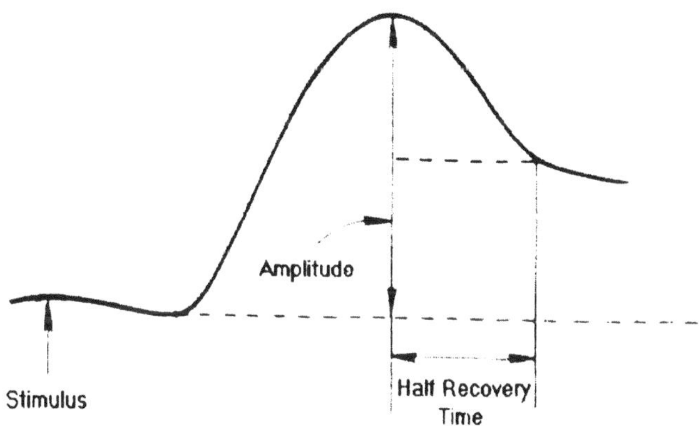

Figure 6. Representation of featured skin conductance response components (i.e., amplitude and half-recovery time). Adapted from Dawson, Schell, & Filion (2000) with permission of the authors and Cambridge University Press.

Studies on the relationship between skin conductance and antisocial behavior have been conducted across a number of community-based and institutionalized populations, although the majority of research has been conducted with a subset of criminals characterized as psychopaths (Raine, 1996).

Skin Conductance Arousal

Resting SCL and NS-SCR are generally interpreted as reflecting sympathetic arousal and it is assumed that antisocial individuals are autonomically underaroused compared to non-antisocial individuals. Although there was early speculation that psychopathic criminals, in particular, would show electrodermal underarousal compared to non-psychopathic criminals, a review of studies conducted up through 1978 (Hare, 1978) found that the hypothesis could only be supported when non-significant SCL results across several studies were pooled.

Similarly, findings for NSRs are inconsistent (Hare, 1978). In a review of the 10 studies on SC arousal conducted between 1979 and 1990, only 2 of 7 analyses on SCL and 3 of 7 analyses on NS-SCR were statistically significant (Raine, 1996). Moreover, the significant findings only applied to non-psychopathic criminals or children with conduct disorder. Although a later study found that behaviorally disordered 11-year-old children with lower SCLs are more likely to be institutionalized at age 13 years, they are not more likely to be arrested (Kruesi *et al.*, 1992). Thus, although there is a possibility that electrodermal underarousal is associated with non-psychopathic antisocial behavior (Buikhuisen, Eurelings-Bontekoe, & Host, 1985; Venables, 1989), the notion that electrodermal underarousal is associated with psychopathy has not met with strong support.

Skin Conductance Orienting

The skin conductance orienting response (SCOR) was originally dubbed by Pavlov as the "investigatory" or "what is it" reflex (Ohman, Hamm, & Hugdahl, 2000). It refers to an increase in SC that occurs when an individual is presented with a novel, neutral stimulus. The SCOR is believed to provide a measure of an individual's attention allocation and level of information processing, with the normative pattern characterized by SCRs to initial stimulus presentations followed by habituation (i.e., gradual decrease and disappearance) of the SCR (Dawson, Schell, & Filion, 2000). When significant findings are observed, they are almost invariably detected for SCOR frequency rather than for SCOR amplitude. This discrepancy is assumed to relate to the fact that amplitude is affected by size and number of sweat glands, which are unrelated to actual functioning of the ANS (Raine, 1996).

Among antisocial individuals, a reduced SCOR is most consistently observed among those with co-existing schizoid or schizotypal personality traits; it is not well-replicated in studies with non-schizoid/schizotypal antisocial individuals. For example, 4 out of 4 studies found significantly lower SCOR in schizoid/schizotypal antisocial groups (Blackburn, 1979; Raine 1987; Raine, Bihrle, Venables, Mednick, & Pollack, 1999; Raine & Venables, 1984), whereas only 2 out of 6 studies found significant SCOR deficits in non-schizoid/schizotypal antisocial groups (Damasio, Tranel, & Damasio, 1990; Raine *et al.*, 1990). Interestingly, findings on reduced SCOR in non-criminal schizophrenics and schizotypals are also mixed (Gruzelier & Raine, 1994), suggesting that SCOR deficits may be most likely to occur when both disorders are present.

With regard to the positive findings from the two non-schizotypal studies, the results of Damasio, Tranel, and Damasio (1990) are from a group of frontal brain lesion patients exhibiting pseudo-sociopathic behavior. Thus, the unique characteristics of this group may limit generalizability to developmental antisocial or psychopathic populations. The results of Raine, Venables, and Williams (1990) are from a longitudinal study that finds an association between reduced SCOR at age 15 in non-institutionalized males and criminal conviction at age 24. Given the general failure of studies to find SCOR deficits in non-schizotypal antisocials, it is possible that this was a spurious finding.

In contrast, a preliminary research base offers a more compelling suggestion that *enhanced* SCOR is a protective factor against the development of criminal behavior in at-risk individuals. Non-criminal males at risk for engaging in criminal behavior by virtue of having a criminal father exhibit higher skin conductance (and heart rate) orienting reactivity compared to criminal males with criminal fathers and controls (i.e., criminal and non-criminal males without criminal fathers; Brennan *et al.*, 1997). In addition, antisocial adolescent males who avoid criminal convictions through age 29 years have higher skin conductance orienting (as well as higher resting skin conductance and heart rate; better skin conductance conditioning; and faster SC recovery times) than antisocial adolescents who grow up to be convicted (Raine, Venables, & Williams, 1995, 1996). It is possible that increased orienting (and other indicators of increased psychophysiological arousal), reflect enhanced information-processing and attentional capacities that help divert at-risk individuals from continuing on an antisocial path (Raine *et al.*, 1995).

Responses to Aversive Stimuli

Another early hypothesis attempting to account for (recidivistic) antisocial behavior is that antisocial individuals, relative to non-antisocial individuals, fail to respond to punishment. This hypothesis has typically been operationalized by measuring SCRs to aversive stimuli such as a loud tone. A review of studies on psychopaths prior to 1978 found evidence that psychopaths consistently show lower SCRs to aversive tones than do non-psychopaths (Hare, 1978). An additional study found that psychopaths fail to show SCRs when witnessing a confederate model pain in response to an electric shock (no shock was actually administered to the confederate; House & Milligan, 1976). Similarly, non-anxious criminals—regardless of psychopathy level—exhibit lower SCRs to modeled pain than do high-anxious criminals.

In contrast, a review (Raine, 1996) of studies published after 1979 failed to substantiate these earlier findings, both when psychopaths were compared to non-psychopaths (4 out of 4 studies non-significant) and when non-psychopathic antisocial individuals were compared to normal controls (4 out of 6 studies non-significant). Thus, recent research has not supported the traditional notion that antisocial individuals are autonomically under-responsive to punishment (Raine, 1996).

Half-Recovery Time

Findings comparing antisocial individuals and controls on half-recovery time, or the amount of time it takes a SCR to return to half its peak size, are fairly consistent (see Figure 6 for a pictorial representation of half-recovery time). Two reviews concluded that SC half-recovery time to novel and aversive tones is significantly longer in psychopaths (Hare, 1978) and criminals (Raine, 1996) than in controls. However, although the findings themselves are reliable, there is disagreement over how to interpret them. Mednick asserts that half-recovery time represents the time it takes for a person's fear to dissipate. He posits that fast fear dissipation—as reflected by short half-recovery time—provides strong reinforcement for avoiding the same behavior in the future. Thus, antisocial individuals, by virtue of their slow half-recovery times, are less likely to refrain from criminal activity due to faulty passive avoidance learning (see below).

Alternatively, Venables (1975) proposes a central processing theory stating that either a decreased excitatory influence of the amygdala or an increased inhibitory influence of the hippocampus serves as a "sensory gate." Thus, slow half-recovery time reflects a closed stance to environmental information that allows one to screen out the aversive nature of punishment. In contrast to Mednick and Venables, however, Fowles (1993) argues that SC half-recovery is a function of amplitude. As such, slow half-recovery merely reflects low SC amplitude and not a particular cognitive process. Given these discrepant interpretations, future research should assess whether SC half-recovery time explains any variance in antisocial behavior over and above SC amplitude.

Passive Avoidance Learning

It is generally thought that avoidance learning occurs in two stages (i.e., classical and operant conditioning; Hare, 1970). First, both behavioral and emotional responses to an aversive stimulus must be classically conditioned. In this paradigm, an unconditioned stimulus (US; e.g., electric shock) elicits an unconditioned (behavioral) response (UR; e.g., withdrawal from shock). When a conditioned stimulus (CS, e.g., a neutral cue) is presented shortly before the shock, after repeated presentations the neutral cue will eventually elicit the behavioral withdrawal independent of the electric shock (i.e., the conditioned response). Because the neutral CS has become associated with an aversive event, the CS is now perceived as negative and elicits fear (i.e., a conditioned emotional response), which is typically measured through autonomic nervous system responding (Zimbardo, 1988). It is believed that the degree of anxiety or fear a person experiences is inversely related to the duration of the delay prior to the anticipated punishment (Hare, 1970).

Once fear of punishment has been conditioned, the person then begins an operant conditioning phase. That is, the person must engage in appropriate and effective fear-reducing behavior when confronted with fear-arousing cues. The reduction in fear accompanying this behavior should encourage the individual to repeat the avoidance behavior when presented with the threatening cues again in the future (Hare, 1970). The likelihood of initiating avoidance of a previously punished response is dependent upon the conditioned fear associated with the cues for that response.

Most of the research on avoidance learning has been conducted with psychopaths, and studies converge in their findings that psychopaths, compared to non-psychopaths, exhibit smaller skin conductance responses while anticipating punishment (Hare, 1965; Lippert & Senter, 1966; Tharp, Maltzman, Syndulko, & Ziskind, 1980) or another person's pain (Aniskiewicz, 1979). Research also suggests that psychopaths are not motivated to learn how to complete maze tasks, even if learning facilitates avoidance of electric shock (Lykken in Hare, 1970; Schachter & Latane, 1964; Schmauk, 1970).

It is generally assumed that psychopaths' impaired avoidance learning is due to deficient sympathetic nervous system activity. Consistent with this hypothesis, psychopaths' ability to learn shock avoidance becomes comparable to that of the controls after receiving an injection of epinephrine, a sympathomimetic agent (Schachter & Latane, 1964). Some studies also find that poor SC conditioning is only observed in—or is especially strong in—low anxious psychopaths (e.g., Aniskiewicz, 1979). In addition, when the maze tasks described above consist of a punishment that is presumably more personally meaningful to the psychopaths (i.e., monetary), they learn to perform the task as well as the controls (Arnett, Howland, Smith, & Newman, 1993; Schmauk, 1970).[1] Thus, it appears that physiological, personality, or motivational factors that augment sympathetic responsivity improve a psychopathic individual's ability to learn avoidance behavior.

Although most research has looked at the SC conditioning of psychopaths, at least two have studied non-psychopathic antisocial individuals. Interestingly, both of these studies fail to find a general relationship between antisocial behavior and poor SC conditioning. Instead, poorer SC conditioning in adult criminals (Hemming, 1981) and conduct-disordered children (Raine & Venables, 1981) is specifically observed in participants who come from relatively good social backgrounds. It has been suggested that while both social and psychophysiological factors are important to the development of antisocial behavior, the interaction of the two may be critical in determining whether the individual is conditioned into an antisocial lifestyle (Raine, 1988). That is, individuals with intact conditioning who are raised in a disruptive environment may be conditioned into an antisocial lifestyle, whereas individuals from prosocial backgrounds experience a relatively greater biological "push" into antisocial behavior (e.g., by failing to condition prosocial behaviors or fear of punishment, and/or by experiencing psychophysiological underarousal that predisposes to stimulation-seeking; Raine, 1997).

Summary of Electrodermal Research

Overall, the electrodermal research suggests that psychopathic criminals, relative to non-psychopathic criminals, exhibit slower half-recovery time and poorer avoidance learning of punishment. In contrast, research with non-psychopathic community samples suggests that antisocial individuals are characterized by electrodermal underarousal compared to their non-antisocial peers. Moreover, among these non-psychopathic antisocials, poor avoidance learning only appears to be characteristic of those who come from advantageous backgrounds. Finally, there is some suggestion that an enhanced SC orienting response may serve as a protective factor for individuals at risk for developing criminal behavior.

Cardiovascular Functioning and Criminal Behavior

Overview

Cardiovascular functioning refers to the functioning of the heart and blood circulatory system. Although one may obtain a variety of cardiovascular measures, the measure most relevant to research on antisocial behavior is heart rate. Heart rate can be measured as a tonic or resting measure (e.g., how many times the heart beats per minute when a person is sitting still and presumably not under stress) or as a phasic measure (e.g., how heart rate changes in response to a stimulus). Although the concept of heart rate seems simple, maintenance and changes in heart rate actually result from complex interactions between the central and autonomic nervous systems.

Both the SNS and PNS regulate heart rate, with the SNS increasing and the PNS decreasing heart rate, respectively (Spence & Mason, 1992). ANS activation as measured by heart rate may therefore reflect either excitatory or inhibitory activity depending on the relative degree of SNS and PNS input. Due to this dual innervation, heart rate has traditionally been used as a general index of autonomic arousal. However, because vagal tone is an important concept in the literature on the psychophysiology of antisocial behavior, it deserves special mention here. Vagal tone refers to the degree to which the vagus nerve of the PNS affects the heart with respect to an inhibitory process. Although computational techniques (i.e., heart rate variability, heart rate variation during paced respiration) to assess vagal tone have been used to estimate parasympathetic control of the heart, these methods are not truly independent of SNS influences (Brownley, Hurwitz, & Schneiderman, 2000). More recently, however, techniques such as spectral analysis have allowed researchers to isolate parasympathetic cardiac activity and directly assess vagal tone.

Heart Rate Reactivity

The principal phasic heart rate responses studied are deceleration and acceleration. Cardiac deceleratory responses occur when the heart briefly slows down in response to the presentation of a neutral stimulus, and acceleratory responses occur when the heart briefly accelerates in response to a (typically aversive) stimulus. Large deceleratory responses are thought to reflect greater attentional processing and orienting, and large acceleratory responses are thought to reflect a defensive response (Raine, 1993). Despite the theoretical relevance of phasic heart rate to criminal behavior, few such studies have been conducted. This is largely due to methodological considerations: phasic heart rate, relative to tonic heart rate, is much more susceptible to methodological problems such as movement artifact and respiration changes elicited by stimulus presentation.

Although the overall research on phasic heart rate and psychopathy is contradictory (Hare, 1978; Raine, 1996), studies looking at reactivity during anticipation of an aversive stimulus in select conditions have produced fairly consistent results. In situations where allocation of attention during an experimental manipulation is *un*controllable, psychopaths demonstrate significant anticipatory heart rate increases compared to non-psychopathic controls (Hare, 1982; Hare & Craigen, 1974; Hare, Frazelle, & Cox, 1978; Ogloff & Wong, 1990). Moreover, these increases in heart rate are often observed in studies finding non-significant differences in skin conductance activity (i.e., Hare, 1982; Hare *et al.*, 1978; Ogloff & Wong, 1990; but see Tharp, Maltzman, Syndulko, & Ziskind, 1980, for an exception with non-institutionalized, psychopathic gamblers).

In contrast, during laboratory tasks where participants have the option of focusing on relatively pleasant stimuli, psychopaths demonstrate cardiovascular underarousal. For example, when participants are simply given the option of listening to an interesting distraction task during anticipation of the aversive stimulus, psychopaths exhibit significantly lower heart rate and non-specific skin conductance responses than the non-psychopaths (Hare, 1982). This general autonomic underarousal in psychopaths is also observed in a computerized Go/No-Go task where participants face monetary punishment in the context of pursuing monetary rewards (Arnett, Howland, Smith, & Newman, 1993). The behavioral and autonomic responding patterns of the two groups in the Arnett *et al.* study further suggest that psychopaths are more cardiovascularly responsive to reward, whereas non-psychopaths are more cardiovascularly responsive to punishment.

Summary

Overall, it appears that psychopaths show increased anticipatory heart rate but decreased skin conductance response when the punishment is unavoidable and salient. This increase in heart rate has been interpreted to represent an internal, physiological coping mechanism that allows psychopaths to tune out cues signaling exposure to aversive stimuli, and low skin conductance has been interpreted to indicate successful coping (e.g., Fowles, 1980; Hare, 1982). Thus, as a result of missing premonitory cues, psychopaths are less "conditionable" to punishment. However, lack of appropriate control groups and failure to obtain concurrent self-report mood ratings make this interpretation of the data questionable. It is possible, for example, that increased heart rate is a physiological representation of excitement or anxiety. It is also possible that Hare's non-psychopathic controls are actually showing the abnormal psychophysiological profile by failing to demonstrate cardiac acceleration during the unavoidable stressor. Nevertheless, psychopaths do exhibit autonomic underarousal in anticipation of, or in response to, punishment when they are distracted by something personally interesting (e.g., monetary rewards). Moreover, they appear to be hyper-responsive to rewards. Collectively, these findings suggest that psychopaths condition more easily to reward and fail to condition to punishment unless they are sufficiently motivated.

Resting Heart Rate

In contrast to phasic heart rate, tonic heart rate is frequently assessed in various adult and child samples and is consistently lower in antisocial individuals relative to controls (Raine, 1996). In a meta-analysis pooling 29 independent samples from studies conducted up through 1992, the average effect size was 0.56 (Raine, 1996), indicating a moderate relationship between the two variables (Cohen, 1988). Moreover, unlike any other psychophysiological correlate reviewed so far (i.e., EEG, SC), no other psychiatric condition is linked to low resting heart rate. In fact, other psychiatric conditions such as alcoholism, depression, schizophrenia, and anxiety are actually associated with higher resting heart rate (Raine, under review).

Interestingly, observed gender differences in antisocial behavior also correspond to observed gender differences in lowered heart rate, with males having significantly lower resting heart rates (and greater antisocial behavior) than females (Voors, Webber, & Berenson, 1982). Furthermore, this gender difference in heart rate emerges at age 3 years, just before gender differences in antisocial, aggressive behavior begin to emerge at age 4 years (Shaw & Winslow, 1997). In addition, the magnitude of the heart rate-antisocial behavior association *within* gender (i.e., non-antisocial versus antisocial females, or non-antisocial versus antisocial males) is the same *across* gender (i.e., *d*s = .50 and .63 for males and females, respectively).

Importantly, these studies have ruled out possible confounds such as physical size and development (Farrington, 1997; Raine, Venables, & Mednick, 1997; Wadsworth, 1976); exercise and sports activity level (Farrington 1997; Wadsworth, 1976); excess motor activity and inattention (Farrington, 1997; Raine *et al.*, 1997); substance abuse (Raine *et al.*, 1997); intellect and academic achievement (Farrington, 1997; Raine *et al.*, 1990); and various forms of social and familial adversity (Farrington, 1997; Raine *et al.*, 1990; Wadsworth, 1976). The low heart rate-antisocial behavior relationship has also been found in six countries, thus attesting to its robustness across cultural contexts (Canada, e.g., Mezzacappa *et al.*, 1997; England, e.g., Farrington, 1987; Germany, e.g., Schmeck & Poustra, 1993; Mauritius, e.g., Raine *et al.*, 1997; New Zealand, e.g., Moffitt & Caspi, in press; United States, e.g., Rogeness, Cepeda, Macedo, Fischer, & Harris, 1990). In addition, longitudinal studies demonstrate that lower resting heart rate precedes antisocial behavior (e.g., Kindlon *et al.*, 1995; Raine *et al.*, 1990, 1997; Wadsworth, 1976), thus attenuating the argument that an antisocial lifestyle results in lowered heart rate.

Consistent with some of the SC findings, heart rate also appears to interact with psychosocial risk factors in predisposing individuals to antisocial behavior. Boys with low resting heart rates are more likely to become violent adult offenders if they have a poor relationship with their parents or if they come from a large family (Farrington, 1997). Boys with low heart rate are also more aggressive if their mother was a teenager when pregnant, if they come from a low SES family, or if they were separated from a parent by age 10 (Farrington, 1997).

Summary of Resting Heart Rate: A Reflection of Vagal Tone Activity?

Low resting heart rate is perhaps one of the best replicated biological markers of antisocial behavior, particularly among children (Raine, under review). Several theories have been generated to account for this robust finding, one of which centers on vagal tone. It has been argued that measures of resting autonomic function taken in a laboratory setting are not actually true measures of rest. Rather, due to the novelty of the situation and the anticipation of the upcoming experimental tasks, these "resting" measures actually capture a person's autonomic activity during mildly stressful situations. Drawing upon this speculation, Raine and Venables (1984) initially suggested that the low heart rate recorded in antisocial individuals reflects increased vagal tone (or parasympathetic activity), thus signaling an enhanced passive coping response to mildly stressful situations. However, data collected since then do not support this hypothesis. Using spectral analysis—a more accurate measure of parasympathetic activity than the computational techniques used by Raine and Venables—Mezzacappa *et al.* (1997) not only find that antisocial 15-year-old boys have lower resting heart rates than non-antisocial boys, but the antisocial boys also have *reduced*, not increased, vagal functioning. Should these findings be replicated in the future, it would suggest that the low resting heart rate-antisocial relationship is driven by particularly strong sympathetic underarousal, which converges with findings of disrupted electrodermal function in antisocials.

Integration of Psychophysiology and Antisocial Behavior

Neuroanatomy of Antisocial Behavior and Autonomic Function

As noted above, the key brain structures involved in modulating autonomic activity in response to environmental stimuli include the orbital and dorsolateral PFCs, the anterior cingulate, insular cortex, hypothalamus, and amygdala. Interestingly, human brain lesion studies, human functional brain-imaging studies, and experimental animal research implicate each of these areas in antisocial behavior. For example, EEG and ERP studies consistently find that (1) antisocial individuals, particularly alcoholic and repetitively violent individuals, exhibit significant frontal and temporal abnormalities, and that (2) psychopaths exhibit widespread cortical abnormalities during information processing. Functional imaging studies find similar evidence of abnormal frontal (Amen, Stubblefield, Carmichael, & Thisted, 1996; Raine, Buchsbaum *et al.*, 1994; Raine, Meloy *et al.*, 1998; Raine, Stoddard, Bihrle, & Buchsbaum, 1998), temporal, and/or subcortical activation in violent offenders or neuropsychiatric patients relative to non-violent controls (Amen *et al.*, 1996; Critchley, Simmons, *et al.*, 2000; Lee *et al.*, 1998; Raine, Park *et al.*, 2001; Raine, Stoddard *et al.*, 1998; Seidenwurm, Pounds, Globus, & Valk, 1997). In addition, during an aversive conditioning paradigm, antisocial individuals—relative to non-antisocial controls—show atypical activation patterns in the dorsolateral prefrontal cortex and amygdala (Schneider *et al.*, 2000), which is consistent with prior behavioral and electrodermal findings.

Due to limitations in technology, research with humans has not systematically examined whether a relationship exists between the hypothalamus and aggressive behavior. However, an extensive body of research with cats demonstrates that electrical stimulation of, or lesions to, specific areas of the hypothalamus consistently facilitate or delay predatory attack behavior, defensive aggression, displays of anger, and aggressive responding to neutral stimuli (Watson, Edinger, & Siegel, 1983). In addition, the medial and lateral hypothalamus appear to modulate defensive and predatory aggression by integrating information from cortical-limbic structures such as the prefrontal cortex, cingulate gyrus, ventral and dorsal hippocampus, amygdala, and septal area (Mirsky & Siegel, 1994).

In addition to the involvement of these particular brain structures, right-hemisphere deficits, in particular, may underlie the autonomic function–antisocial behavior relationship. Lesion and intracarotid amobarbital studies demonstrate that decreased right-hemisphere functioning reduces heart rate (Yokoyama, Jennings, Ackles, Hood, & Boller, 1987; Zamrini *et al.*, 1990). Similarly, control of human electrodermal regulation is predominant in the right hemisphere (Tranel & Damasio, 1994; Zahn, Grafman, & Tranel, 1999), with some suggestion that the lateralization is particularly strong for emotionally relevant stimuli (Zoccolotti, Scabini, & Violani, 1982). In addition, poor right-hemisphere functioning (particularly in the anterior regions) is associated with withdrawal system deficits. According to Davidson and colleagues, the withdrawal system facilitates an individual's retreat from aversive or dangerous situations (Davidson, 1998; Davidson, Eckman, Saron, Senulis, & Friesen, 1990). Thus, reduced right-hemisphere functioning may contribute to the likelihood of a child engaging in dangerous, risky, and/or antisocial behavior. Poor right-hemisphere dysfunction as measured by fMRI (e.g., Raine, Park *et al.*, 2001), computed tomography (e.g., Hucker *et al.*, 1988), neuropsychological tests (e.g., Day & Wong, 1996), EEG (e.g., Evans & Park, 1997), and ERPs (e.g., Drake, Pakalnis, Brown, & Hietter, 1988) is also observed in antisocial and violent populations. Similarly, patients with right hemisphere lesions, compared to those with left hemisphere lesions, show reduced heart rate and skin conductance responses to films depicting negative emotions (Zoccolotti, Caltagirone, Benedetti, & Gainotti, 1986). However, the importance and specificity of right-hemispheric dysfunction both to reduced autonomic responding and elevated aggressive or antisocial behavior is not well understood and should be considered speculative, especially since left-hemisphere dysfunction is also implicated in violence and crime (Raine, 1993).

Central and Autonomic Mechanisms Underlying Antisocial Behavior: The Somatic Marker Hypothesis

Traditionally, neuroanatomical and psychophysiological research have been conducted independently. Recently, however, there have been attempts to link the fields into a coherent model reflecting a biological basis to human aggression (Damasio, 1994; Davidson, Putnam, & Larson, 2000). Damasio (1994) has described a neurophysiological theory to account for poor social decision making observed in antisocial individuals. Central to his theory is an individual's ability to generate emotions, which are manifested as alterations in physiological state and registered in the brain as changes in somatosensory structure activation (Bechara, Damasio, Damasio, & Lee, 1999). Damasio posits that conditioned emotional responses facilitate enhanced decision making by reducing the number of options from which an individual has to choose. The reduction occurs when an individual experiences uneasy feelings that were conditioned during prior negative experiences with that option, and he intuitively or deliberately decides to eliminate it from further consideration. As a result, inappropriate decisions are generally excluded, and cognitive resources—particularly frontal functions such as planning flexibility and working memory—are freed up to evaluate only options that are viable to that individual.

The risk for antisocial behavior is elevated when an individual cannot generate or does not appreciate the significance of somatic markers (see Chapter 7 for an extensive review). Failure to generate somatic markers may occur with damage to the dorsolateral PFC (Grueninger et al., 1965; Tranel & Damasio, 1994), orbital PFC (Critchley, Elliott et al., 2000), amygdala (Bechara, Damasio, Damasio, & Lee, 1999), hypothalamus, and other structures involved in emotion and psychophysiological responses (e.g., cingulate, insula, motor cortex, dorsal pons, midbrain, cerebellum; Damasio, Grabowski et al., 2000; Davidson, Putnam, & Larson, 2000). Furthermore, even if one is able to generate somatic markers, prefrontal ventromedial dysfunction may nevertheless render one unable to utilize this information by interfering with "affective working memory," or the ability to use emotional memory to guide current behavior based on anticipation of future consequences (Bechara et al., 1999; Bechara, Tranel, & Damasio, 2000; Davidson, Jackson, & Kalin, 2000). Clearly, the above review on deficient avoidance learning in developmental psychopaths, and the enhanced autonomic functioning observed in at-risk individuals who avoid crime, fit particularly well within this model. Low resting heart rate in antisocial individuals may similarly reflect disruption of the somatic marker network and consequently increase risk-related behavior. That is, low levels of arousal during mildly stressful psychophysiological test sessions may reflect low levels of fear (Raine, 1993, 1997). This fearlessness may predispose the individual to antisocial behavior by impairing avoidance learning of punishment, whether the punishment is the fear of getting arrested or being injured in a fight (Raine, 1993).

In addition to impairing avoidance learning and risk-related decision making, low resting heart rate may provide a psychophysiological impetus for some individuals to engage in antisocial behavior. Low heart rate, particularly in conjunction with other measures of autonomic (i.e., skin conductance; Kruesi *et al.*, 1992) and central nervous system underarousal (e.g., resting EEG; Raine *et al.*, 1990), may represent an unpleasant physiological state. As a result, antisocials may be prompted to seek stimulation in order to elevate their arousal to an optimal or normal level (Eysenck, 1997; Quay, 1965; Raine *et al.*, 1997). Consistent with this hypothesis, low resting heart rate at age 3 characterizes both stimulation-seeking behavior at age 3 and aggressive behavior at age 11 (Raine *et al.*, 1997; Raine, Reynolds, Venables, Mednick, & Farrington, 1998). In addition, not only do preschool boys with externalizing problems have lower resting heart rate compared to normal preschool boys, but boys with externalizing behavior problems are also more likely to watch videotapes depicting intense anger (El-Sheikh *et al.*, 1994). Thus, antisocial behavior in generally underaroused individuals may not only reflect impairment of the somatic marker system, but may also represent physiological compensation in that committing a burglary, assault, or substance abuse is positively stimulating for some individuals.

CONCLUSIONS

Despite the general convergence of these different biological correlates into a viable neurophysiological model of antisocial behavior, the literature leaves many questions unanswered. Obviously, research needs to assess the links between neuroanatomy, autonomic functions, and antisocial behavior directly. Separate models will also need to be generated for subtypes of antisocial behavior, such as premeditated versus impulsive aggression, or antisocial behavior that is not necessarily criminal (e.g., alcoholism, some forms of drug use). Finally, research will need to address the complex issues of developmental neural plasticity, temperamental differences, the environmental context, and their combined association with regard to antisocial behavior across the lifespan (Davidson, Putnam, & Larson, 2000). Nevertheless, the current body of work demonstrates how the interplay of neurophysiological systems and learning can create a capacity for antisocial behavior, and poses many interesting questions to be addressed in future research.

NOTES

1. Although Schmauk (1970) and Arnett *et al.* (1993) found different results regarding SCRs to monetary punishment (i.e., no group differences versus tendency towards fewer responses in psychopaths, respectively), this may reflect differences in design, which is discussed in the section on "Heart rate reactivity."

REFERENCES

Allen, G.V., & Cechetto, D.F. (1992). Functional and anatomical organization of cardiovascular pressor and depressor sites in the lateral hypothalamic area: I. Descending projections. *The Journal of Comparative Neurology*, **315**, 313-332.

Amen, D.G., Stubblefield, M., Carmichael, B., & Thisted, R. (1996). Brain SPECT findings and aggressiveness. *Annals of Clinical Psychiatry*, **8**, 129-137.

Aniskiewicz, A.S. (1979). Autonomic components of vicarious conditioning and psychopathy. *Journal of Clinical Psychology*, **35**, 60-67.

Arnett, P.A., Howland, E.W., Smith, S.S., & Newman, J.P. (1993). Autonomic responsivity during passive avoidance in incarcerated psychopaths. *Personality and Individual Differences*, **14**, 173-184.

Barratt E.S., Stanford, M.S., Kent, T.A., & Felthous, A. (1997). Neuropsychological and cognitive psychophysiological substrates of impulsive aggression. *Biological Psychiatry*, **41**, 1045-1061.

Bauer, L.O., & Hesselbrock, V.M. (1999). P300 decrements in teenagers with conduct problems: Implications for substance abuse risk and brain development. *Biological Psychiatry*, **46**, 263-272.

Bauer, L.O., O'Connor, S., & Hesselbrock, V.M. (1994). Frontal P300 decrements in antisocial personality disorder. *Alcoholism: Clinical and Experimental Research*, **18**, 1300-1305.

Baving, L., Laucht, M., & Schmidt, M.H. (2000). Oppositional children differ from healthy children in frontal brain activation. *Journal of Abnormal Psychology*, **28**, 267-275.

Bechara, A., Damasio, H., Damasio, A.R., & Lee, G.P. (1999). Different contributions of the human amygdala and ventromedial prefrontal cortex to decision-making. *Journal of Neuroscience*, **19**, 5473-5481.

Bechara, A., Tranel, D., & Damasio, H. (2000). Characterization of the decision-making deficit of patients with ventromedial prefrontal cortex lesions. *Brain*, **123**, 2189-2202.

Bechara, A., Tranel, D., Damasio, H., & Damasio, A.R. (1996). Failure to respond autonomically to anticipated future outcomes following damage to prefrontal cortex. *Cerebral Cortex*, **6**, 215-225.

Blackburn, R. (1979). Psychopathy, delinquency, and crime. In A. Gale & J.A. Edwards (Eds.), *Psychophysiological correlates of human behavior* (Vol. 3, pp. 187-205). London: Academic Press.

Branchley, M.H., Buydens-Branchley, L., & Lieber, C.S. (1988). P3 in alcoholics with disordered regulation of aggression. *Psychiatry Research*, **25**, 49-58.

Brennan, P.A., Raine, A., Schulsinger, F., Kirkegaard-Sorenson, L., Know, J., Hutchings, B., Rosenberg, R., & Mednick, S.A. (1997). Psychophysiological protective factors for male subjects at high risk for criminal behavior. *American Journal of Psychiatry*, **154**, 853-855.

Brownley, K.A., Hurwitz, B.E., & Schneiderman, N. (2000). Cardiovascular psychophysiology. In J.T. Cacioppo, L.G. Tassinary, & G.G. Berntson (Eds.), *Handbook of psychophysiology* (2nd ed., pp. 224-264). New York: Cambridge University Press.

Buikhuisen, W., Eurelings-Bontekoe, E.H.M., & Host, K.B. (1985). Crime and recovery time: Mednick revisited. *International Journal of Law and Psychiatry*, **12**, 29-40.

Cacioppo, J.T., & Tassinary, L.G. (1990). Psychophysiology and psychophysiological inference. In J.T. Cacioppo & L.G. Tassinary (Eds.),

Principles of psychophysiology: Physical, social, and inferential elements (pp. 3-33). New York: Cambridge University Press.

Cohen, J. (1988). *Statistical power for the social sciences* (2nd ed.). Hillsdale, NJ: Erlbaum.

Convit, A., Czobor, P., & Volavka, J. (1991). Lateralized abnormality in the EEG of persistently violent psychiatric inpatients. *Biological Psychiatry*, **30**, 363-370.

Costa, L., Bauer, L.O., Kuperman, S., Porjesz, B., O'Connor, S., Hesselbrock, V.M., Rohrbaugh, J., & Begleiter, H. (2000). Frontal P300 decrements, alcohol dependence, and antisocial personality disorder. *Biological Psychiatry*, **47**, 1064-1071.

Critchley, H.D., Corfield, D.R., Chandler, M.P., Mathias, C.J., & Dolan, R.J. (2000). Cerebral correlates of autonomic cardiovascular arousal: A functional neuroimaging investigation of humans. *Journal of Physiology*, **523**, 259-270.

Critchley, H.D., Elliott, R., Mathias, C.J., & Dolan, R.J. (2000). Neural activity relating to generation and representation of galvanic skin conductance responses: A functional magnetic resonance imaging study. *Journal of Neuroscience*, **20**, 3033-3040.

Critchley, H.D., Simmons, A., Daly, E.M., Russell, A., van Amelsvoort, T., Robertson, D.M., Glover, A., Murphy, D.G.M. (2000). Prefrontal and medial temporal correlates of repetitive violence to self and others. *Biological Psychiatry*, **47**, 928-934.

Cummings, J.L. (1995). Anatomic and behavioral aspects of frontal-subcortical circuits. In J. Grafman, K.J. Holyoak, & F. Boller (Eds.), *Structure and functions of the human prefrontal cortex. Annals of the New York Academy of Sciences*, **769**, 1 13.

Damasio, A.R. (1994). *Descartes' error: Emotion, reason, and the human brain.* New York: Grosset/Putnam.

Damasio, A.R., & Anderson, S.W. (1993). The frontal lobes. In K.M. Heilman & E. Valenstein (Eds.), *Clinical neuropsychology* (3rd ed., pp. 409-460). New York: Oxford University Press.

Damasio, A.R., Grabowski, T.J., Bechara, A., Damasio, H., Ponto, L.L.B., Parvizi, J., & Hichwa, R.D. (2000). Subcortical and cortical brain activity during the feeling of self-generated emotions. *Nature Neuroscience*, **3**, 1049-1056.

Damasio, A.R., Tranel, D., & Damasio, H. (1990). Individuals with psychopathic behavior caused by frontal damage fail to respond autonomically to social stimuli. *Behavioral and Brain Research*, **41**, 81- 94.

Davidson, R.J. (1998). Anterior electrophysiological asymmetries, emotion, and depression: Conceptual and methodological conundrums. *Psychophysiology*, **35**, 607-614.

Davidson, R.J., Eckman, P., Saron, C.D., Senulis, J.A., & Friesen, W.V. (1990). Approach-withdrawal and cerebral asymmetry: Emotional expression and brain physiology I. *Journal of Personality and Social Psychology*, **58**, 330-341.

Davidson, R.J., Jackson, D.C., & Kalin, N.H. (2000). Emotion, plasticity, context and regulation: Perspectives from affective neuroscience. *Psychological Bulletin*, **126**, 890-909.

Davidson, R.J., Putnam, K.M., & Larson, C.L. (2000). Dysfunction in the neural circuitry of emotion regulation: A possible prelude to violence. *Science*, **289**, 591-594.

Dawson, M.E., Schell, A.M., & Filion, D.L. (2000). The electrodermal system. In J.T. Cacioppo, L.G. Tassinary, & G.G. Berntson (Eds.), *Handbook of psychophysiology* (2nd ed., pp. 200-223). New York: Cambridge University Press.

Day, R., & Wong, S. (1996). Anomalous perceptual asymmetries for negative emotional stimuli in the psychopath. *Journal of Abnormal Psychology*, **105**, 648-652.

Deckel, A.W., Hesselbrock, V., & Bauer, L. (1996). Antisocial personality disorder, delinquency, and frontal brain dysfunctioning: EEG and neuropsychological findings. *Journal of Clinical Psychology*, **52**, 639-650.

DeArmond, S.J., Fusco, M.M., & Dewey, M.M. (1989). *Structure of the human brain: A photographic atlas* (2nd ed.). New York: Oxford University Press.

Drake, M.E., Hietter, S.A., & Pakalnis, A. (1992). EEG and evoked potentials in episodic-dyscontrol syndrome. *Neuropsychobiology*, **26**, 125-128.

Drake, M.E., Pakalnis, A., Brown, M.E., & Hietter, S.A. (1988). Auditory event related potentials in violent and nonviolent prisoners. *European Archives of Psychiatry & Neurological Sciences*, **238**, 7-10.

El-Sheikh, M., Ballard, M., & Cummings, E.M. (1994). Individual differences in preschoolers' physiological and verbal responses to videotaped angry interactions. *Journal of Abnormal Child Psychology*, **22**, 303-320.

Evans, J.R., & Park, N.S. (1997). Quantitative EEG findings among men convicted of murder. *Journal of Neurotherapy*, **2**, 31-39.

Eysenck, H.J. (1997). Personality and the biosocial model of antisocial and criminal behavior. In A. Raine, P. Brennan, D.P. Farrington, & S.A. Mednick (Eds.)., *Biosocial bases of violence* (pp. 21-38). New York: Plenum.

Fabiani, M., Gratton, G., & Coles, M.G.H. (2000). Event-related brain potentials: Methods, theory, and applications. In J.T. Cacioppo, L.G. Tassinary, & G.G. Berntson (Eds.), *Handbook of psychophysiology* (2nd ed., pp. 53-84). New York: Cambridge University Press.

Farrington, D.P. (1987). Implications of biological findings for criminological research. In S.A. Mednick, T.E. Moffitt, & S.A. Stack (Eds.), *The causes of crime: New biological approaches* (pp. 42-64). New York: Cambridge University Press.

Farrington, D.P. (1997). The relationship between low resting heart rate and violence. In A. Raine, P.A. Brennan, D.P. Farrington, & S.A. Mednick (Eds.), *Biosocial bases of violence* (pp. 89-106). New York: Plenum.

Fishbein, D.H., Herning, R.I., Pickworth, W.B., Haertzen, C.A., Hickey, J.E., & Jaffe, J.H. (1989). EEG and brainstem auditory evoked response potentials in adult male drug abusers with self-reported histories of aggressive behavior. *Biological Psychiatry*, **26**, 595-611.

Fowles, D.C. (1980). The three arousal model: Implications of Gray's two-factor learning theory for heart rate, electrodermal activity, and psychopathy. *Psychophysiology,* **17**, 87-104.

Fowles, D.C. (1993). Electrodermal activity and antisocial behavior: Empirical findings and theoretical issues. In J.C. Roy, W. Boucsein, D.C. Fowles, & J.H. Gruzelier (Eds.), *Progress in electrodermal research* (pp. 223-238). New York: Plenum.

Fox, N.A., Schmidt, L.A., Calkins, S.D., Rubin, K.H., & Coplan, R.J. (1996). The role of frontal activation in the regulation and dysregulation of social behavior during the preschool years. *Development and Psychopathology*, **8**, 89-102.

Frederickson, M., Furmark, T., Olsson, M.T., Fischer, H., Anderson, J., & Langstrom, B. (1998). Functional neuroanatomical correlates of electrodermal activity: A positron emission tomographic study. *Psychophysiology,* **35**, 179-185.

Gatzke-Kopp, L., Raine, A., Buchsbaum, M.S., & LaCasse, L. (in press). Temporal lobe EEG deficits in murderers not detected by PET. *Journal of Neuropsychiatry and Clinical Neurosciences.*

Gerstle, J.E., Mathias, C.W., & Stanford, M.S. (1998). Auditory P300 and self-reported impulsive aggression. *Progress in Neuropsychopharmacology and Biological Psychiatry*, **22**, 575-583.

Grueninger, W.E., Kimble, D.P., Grueninger, J., & Levine, S. (1965). GSR and corticosteroid response in monkeys with frontal ablations. *Neuropsychologia,* **3**, 205-216.

Gruzelier, J., & Raine, A. (1994). Bilateral electrodermal activity and cerebral mechanisms in syndromes of schizophrenia and the schizotypal personality. *International Journal of Psychophysiology*, **16**, 1-16.

Hare, R.D. (1965). Temporal gradient of fear arousal in psychopaths. *Journal of Abnormal Psychology,* **70**, 442-445.

Hare, R.D. (1970). *Psychopathy: Theory and research.* New York: John Wiley & Sons.

Hare, R.D. (1978). Electrodermal and cardiovascular correlates of psychopathy. In R.D. Hare, & D. Schalling (Eds.), *Psychopathic behavior: Approaches to research* (pp. 107-144). Chichester: John Wiley & Sons.

Hare, R.D. (1982). Psychopathy and physiological activity during anticipation of an aversive stimulus in a distraction paradigm, *Psychophysiology*, **19**, 266-271.

Hare, R.D. (1993). *Without conscience: The disturbing world of the psychopaths among us*. New York: Pocket Books.

Hare, R.D., & Craigen, D. (1974). Psychopathy and physiological activity in a mixed-motive game situation. *Psychophysiology*, **11**, 197-206.

Hare, R.D., Frazelle, J., & Cox, D.N. (1978). Psychopathy and physiological responses to threat of an aversive stimulus. *Psychophysiology*, **15**, 165-172.

Hemming, J.H. (1981). Electrodermal indices in a selected prison sample and students. *Personality and Individual Differences*, **2**, 37-46.

Hill, D., & Pond, D.A. (1952). Reflections on 100 capital cases submitted for electroencephalography. *Journal of Mental Science*, **98**, 23-43.

House, T.H., & Milligan, W.L. (1976). Autonomic responses to modeled distress in prison psychopaths. *Journal of Personality and Social Psychology*, **34**, 556-560.

Hucker, S., Langevin, R., Wortzman, G., Dickey, R., Bain, J., Handy, L., Chambers, J., & Wright, S. (1988). Cerebral damage and dysfunction in sexually aggressive men. *Annals of Sex Research*, **1**, 33-47.

Hugdahl, K. (1995). *Psychophysiology: The mind-body perspective*. Cambridge, MA: Harvard University Press.

Hugdahl, K. (1998). Cortical control of human classical conditioning: Autonomic and positron emission tomography data. *Psychophysiology*, **35**, 170-178.

Jutai, J.W., Hare, R.D., & Connolly, J.F. (1987). Psychopathy and event-related brain potentials (ERPs) associated with attention to speech stimuli. *Personality and Individual Differences*, **8**, 175-184.

Kiehl, K.A., Hare, R.D., Liddle, P.F., & McDonald, J.J. (1999). Reduced P300 responses in criminal psychopaths during a visual oddball task. *Biological Psychiatry*, **45**, 1498-1507.

Kiehl, K.A., Hare, R.D., McDonald, J.J., & Brink, J. (1999). Semantic and affective processing in psychopaths: An event-related potential (ERP) study. *Psychophysiology*, **36**, 765-774.

Kiehl, K.A., Smith, A.M., Hare, R.D., & Liddle, P.F. (2000). An event-related potential (ERP) investigation of response inhibition in schizophrenia and psychopathy. *Biological Psychiatry*, **48**, 210-221.

Kindlon, D.J., Tremblay, R.E., Mezzacappa, E., Earls, F., Laurent, D., & Schaal, B. (1995). Longitudinal patterns of heart rate and fighting behavior in 9- through 12-year-old boys. *Journal of the American Academy of Child & Adolescent Psychiatry*, **34**, 371-377.

Kruesi, M.J.P., Hibbs, E., Zahn, T.P., Keysor, C.S., Hamburger, S., Bartko, J.J., & Rapoport, J.L. (1992). A two year prospective follow-up study of children and adolescents with disruptive behavior disorders: Prediction with CSF 5-HIAA, HVA, and autonomic measures? *Archives of General Psychiatry*, **49**, 429-435.

Lahey, B.B., McBurnett, K., Loeber, R., & Hart, E.L. (1995). Psychobiology. In G. Sholevar (Ed.), *Conduct disorders in children and adolescents* (pp. 27-44). Washington, DC: American Psychiatric Press.

LaPierre, D., Braun, C.M.J., & Hodgins, S. (1995). Ventral frontal deficits in psychopathy: Neuropsychological test findings. *Neuropsychologia*, **33**, 139-151.

Larsen, P.B., Schneiderman, N., & Pasin, R.D. (1986). Physiological bases of cardiovascular psychophysiology. In M.G.H. Coles, E. Donchin, & S.W. Porges (Eds.), *Psychophysiology: Systems, processes, and applications* (pp. 122-165). New York: Guilford.

Lee, G.P., Bechara, A., Adolphs, R., Arena, J., Meador, K.J., Loring, D.W., & Smith, J.R. (1998). Clinical and physiological effects of stereotaxic bilateral amygdalotomy for intractable aggression. *Journal of Neuropsychiatry & Clinical Neurosciences*, **10**, 413-420.

Lippert, W.W., & Senter, R.J. (1966). Electrodermal responses in the psychopath. *Psychonomic Science*, **4**, 25-26.

Mark, V.H., & Ervin, F.R. (1970). *Violence and the brain.* New York: Harper & Row.

Mednick, S.A., Volavka, J., Gabrielli, W.F., & Itil, T.M. (1981). EEG as a predictor of antisocial behavior. *Criminology*, **19**, 219-229.

Mezzacappa, E., Tremblay, R.E., Kindlon, D., Saul, J.P., Arseneault, L., Seguin, J, Phil, R.O., & Earls, F. (1997). Anxiety, antisocial behavior, and heart rate regulation in adolescent males. *Journal of Child Psychology and Psychiatry*, **38**, 457-469.

Mirsky, A.F, & Siegel, A. (1994). The neurobiology of violence and aggression. In A.J. Reiss, K.A. Miczek, & J.A. Roth (Eds.), *Understanding and preventing violence: Volume 2, biobehavioral influences* (pp. 59-172). Washington, DC: National Academy Press.

Moffitt, T.E., & Caspi, A. (in press). Childhood predictors differentiate life-course persistent and adolescent limited pathways among males and females. *Development and Psychopathology.*

Ogloff, J.R.P., & Wong, S. (1990). Electrodermal and cardiovascular evidence of a coping response in psychopaths. *Criminal Justice and Behavior*, **17**, 231-245.

Ohman, A., Hamm, A., & Hugdahl, K. (2000). Cognition and the autonomic nervous system: Orienting, anticipation, and conditioning. In J.T. Cacioppo, L.G. Tassinary, & G.G. Berntson (Eds.), *Handbook of psychophysiology* (2nd ed., pp. 533-575). New York: Cambridge University Press.

Oppenheimer, S.M., Gelb, A., Girvin, J.P., & Hachinski, V.C. (1992). Cardiovascular effects of human insular cortex stimulation. *Neurology*, **42**, 1727-1732.

Quay, H.C. (1965). Psychopathic personality as pathological stimulation-seeking. *American Journal of Psychiatry*, **122**, 180-183.

Raine. A. (1987). Effect of early environment on electrodermal and cognitive correlates of schizotypy and psychopathy in criminals. *International Journal of Psychophysiology*, **4**, 277-287.

Raine, A. (1988). Antisocial behavior and social psychophysiology. In H.L. Wagner (Ed.), *Social psychophysiology and emotion: Theory and clinical applications* (pp. 231-250). London: John Wiley & Sons

Raine. A. (1989). Evoked potentials and psychopathy. *International Journal of Psychophysiology*, **8**, 1-16.

Raine, A. (1993). *The psychopathology of crime: Criminal behavior as a clinical disorder*. San Diego: Academic Press.

Raine, A. (1996). Autonomic nervous system activity and violence. In D.M. Stoff & R.B. Cairns (Eds.), *Neurobiological approaches to clinical aggression research* (pp. 145-168). Mahwah, NJ: Lawrence Erlbaum.

Raine, A. (1997). Antisocial behavior and psychophysiology: A biosocial perspective and a prefrontal dysfunction hypothesis. In D.M. Stoff, J. Breiling, & J.D. Maser (Eds.), *The handbook of antisocial behavior* (pp. 289-304). New York: John Wiley & Sons.

Raine, A. (under review). The role of prefrontal deficits, low autonomic arousal, and early health factors in the development of antisocial and aggressive behavior. *Journal of Child Psychology and Psychiatry*.

Raine, A., Bihrle, S., Venables, P.H., Mednick, S.A., & Pollack, V. (1999). Skin conductance orienting deficits and increased alcoholism in schizotypal criminals. *Journal of Abnormal Psychology*, **108**, 299-306.

Raine, A., Buchsbaum, M.S., Stanley, J., Lottenberg, S., Abel, L., & Stoddard, J. (1994). Selective reductions in prefrontal glucose metabolism in murderers. *Biological Psychiatry*, **36**, 366-373.

Raine, A., & Lencz, T. (1993). Brain imaging research on electrodermal activity in humans. In W. Boucsein, D.C. Fowles, & J.H. Gruzelier (Eds.), *Progress in electrodermal research* (pp. 115-135). New York: Plenum Press.

Raine, A., Meloy, J.R., Bihrle, S., Stoddard, J., LaCasse, L., & Buchsbaum, M.S. (1998). Reduced prefrontal and increased subcortical brain functioning assessed using positron emission tomography in predatory and affective murderers. *Behavioral Sciences & the Law*, **16**, 319-332.

Raine, A., Park, S., Lencz, T., Bihrle, S., Lacasse, L., Widom, C.S., Al-Dayeh, L., & Singh, M. (2001). Reduced right hemisphere activation in severely abused violent offenders during a working memory task: An fMRI study. *Aggressive Behavior*, **27**, 111-129.

Raine, A., Reynolds, G.P., & Sheard, C. (1991). Neuroanatomical correlates of skin conductance orienting in normal humans: A magnetic resonance imaging study. *Psychophysiology*, **28**, 548-558.

Raine, A., Reynolds, C., Venables, P.H., Mednick, S.A., & Farrington, D.P. (1998). Fearlessness, stimulation-seeking, and large body size at age 3 years as early predispositions to childhood aggression at age 11 years. *Archives of General Psychiatry*, **55**, 745-751.

Raine, A., Stoddard, J., Bihrle, S., & Buchsbaum, M.S. (1998). Prefrontal glucose deficits in murderers lacking psychosocial deprivation. *Neuropsychiatry, Neuropsychology, and Behavioral Neurology*, **11**, 1-7.

Raine, A., & Venables, P.H. (1981). Classical conditioning and socialization – A biosocial interaction? *Personality and Individual Differences*, **2**, 273-283.

Raine, A., & Venables, P.H. (1984). Tonic heart rate level, social class, and antisocial behavior. *Biological Psychiatry*, **18**, 123-132.

Raine, A., & Venables, P.H. (1988). Enhanced P3 evoked potentials during a continuous performance task in psychopaths. *Psychophysiology*, **25**, 30-38.

Raine, A., & Venables, P.H. (1990). Evoked potential augmenting-reducing in psychopaths and criminals with impaired smooth-pursuit eye movements. *Psychiatry Research*, **31**, 85-98.

Raine, A., Venables, P.H., & Mednick, S.A. (1997). Low resting heart rate at age 3 years predisposes to aggression at age 11 years: Findings from the Mauritius Joint Child Health Project. *Journal of the American Academy of Child and Adolescent Psychiatry*, **36**, 1457-1464.

Raine, A., Venables, P.H., & Williams, M. (1990). Relationships between CNS and ANS measures of arousal at age 15 and criminality at age 24. *Archives of General Psychiatry*, **47**, 1003-1007.

Raine, A., Venables, P.H., & Williams, M. (1995). High autonomic arousal and electrodermal orienting at age 15 years as protective factors against criminal behavior at age 29 years. *American Journal of Psychiatry*, **152**, 1595-1600.

Raine, A., Venables, P.H., & Williams, M. (1996). Better autonomic conditioning and faster electrodermal half-recovery time at age 15 years as possible protective factors against crime at age 29 years. *Developmental Psychology*, **32**, 624-630.

Rogeness, G.A., Cepeda, C., Macedo, C.A., Fischer, C., & Harris, W.R. (1990). Differences in heart rate and blood pressure in children with conduct disorder, major depression, and separation anxiety. *Psychiatry Research*, **33**, 199-206.

Schachter, S., & Latane, B. (1964). Crime, cognition, and the autonomic nervous system. *Nebraska Symposium on Motivation*, **12**, 221-275.

Schmauk, F.J. (1970). Punishment, arousal, and avoidance learning in psychopaths. *Journal of Abnormal Psychology*, **76**, 325-335.

Schmeck, K., & Poustra, F. (1993). Psychophysiologische reacktionsmuster und psychische auffalligkeiten im kindesalter. In P. Baumann (Ed.), *Biologische psychiatrie der gegenwart* (pp. 359-364). New York: Springer-Verlag.

Schneider, F., Habel, U., Kessler, C., Posse, S., Grodd, W., & Muller-Gartner, H.W. (2000). Functional imaging of conditioned aversive emotional responses in antisocial personality disorder. *Neuropsychobiology*, **42**, 192-201.

Seidenwurm, D., Pounds, T.R., Globus, A., & Valk, P.E. (1997). Temporal lobe metabolism in violent subjects: Correlation of imaging and neuropsychiatric findings. *American Journal of Neuroradiology*, **18**, 625-631.

Sequeira, H., & Roy, J.C. (1993). Cortical and hypothalamic-limbic control of electrodermal responses. In J.C. Roy, W. Boucsein, D.C. Fowles, & J.H. Gruzelier (Eds.), *Progress in electrodermal research* (pp. 93-114). New York: Plenum

Shaw, D.S., & Winslow, E.B. (1997). Precursors and correlates of antisocial behavior from infancy to preschool. In D.M. Stoff, J. Breiling, & J.D. Maser (Eds.), *Handbook of antisocial behavior* (pp. 148-158). New York: Wiley.

Spence, A.P., & Mason, E.B. (1992). *Human anatomy and physiology* (4th ed.). St. Paul, MN: West Publishing Co.

Tharp, V.K., Maltzman, I., Syndulko, K., & Ziskind, E. (1980). Autonomic activity during anticipation of an aversive tone in noninstitutionalized sociopaths. *Psychophysiology*, **17**, 123-128.

Tranel, D., & Damasio, H. (1989). Intact electrodermal skin conductance responses after bilateral amygdala damage. *Neuropsychologia*, **27**, 381-390.

Tranel, D., & Damasio, H. (1994). Neuroanatomical correlates of electrodermal skin conductance responses. *Psychophysiology*, **31**, 427-438.

Venables, P.H. (1975). Progress in psychophysiology: Some applications in a field of abnormal psychology. In P.H. Venables & M.J. Christie (Eds.), *Research in psychophysiology* (pp. 418-437). London: John Wiley & Sons.

Venables, P.H. (1989). The Emanuel Miller Memorial Lecture 1987: Childhood markers for adult disorders. *Journal of Child Psychology and Psychiatry and Allied Disciplines*, **30**, 347-364.

Volavka, J. (1987). Electroencephalogram among criminals. In S.A. Mednick, T.E. Moffitt, & S. Stack (Eds.), *The causes of crime: New biological approaches* (pp. 137-145). Cambridge: Cambridge University Press.

Volavka, J. (1995). *Neurobiology of violence.* Washington, DC: American Psychiatric Association.

Voors, A.W., Webber, L.S., & Berenson, B.S. (1982). Resting heart rate and pressure rate product of children in a total biracial community: The Bogalusa Heart study. *American Journal of Epidemiology*, **116**, 276-286.

Wadsworth, M.E.J. (1976). Delinquency, pulse rate and early emotional deprivation. *British Journal of Criminology*, **16**, 245-256.

Watson, R.E., Edinger, H.M., & Siegel, A. (1983). An analysis of the mechanisms underlying hippocampal control of hypothalamically-elicited aggression in the cat. *Brain Research*, **269**, 327-345.

Williams, D. (1969). Neural factors related to habitual aggression: Consideration of those differences between those habitually aggressive and others who have committed crimes of violence. *Brain*, **92**, 503-520.

Williamson, S., Harpur, T.J., & Hare, R.D. (1991). Abnormal processing of affective words by psychopaths. *Psychophysiology*, **28**, 260-273.

Williamson, J.W., Nobrega, A.C.L., McColl, R., Mathews, D., Winchester, P., Friberg, L., & Mitchell, J.H. (1997). Activation of the insular cortex during dynamic exercise in humans. *Journal of Physiology*, **503**, 277-283.

Wong, M.T.H., Lumsden, J., Fenton, G.Q., & Fenwick, P.B.C. (1994). Electroencephalography, computed tomography and violence ratings of male patients in a maximum-security mental hospital. *Acta Psychiatrica Scandinavica*, **90**, 97-101.

Yaralian, P.S., & Raine, A. (2000). Biological approaches to crime: Psychophysiology and brain dysfunction. In R. Paternoster & R. Bachman (Eds.), *Explaining criminals and crime: Essays in contemporary criminological theory* (pp. 57-72). Los Angeles, CA: Roxbury Publishing Co.

Yokoyama, K., Jennings, R., Ackles, P., Hood, P., & Boller, F. (1987). Lack of heart rate changes during an attention-demanding task after right hemisphere lesions. *Neurology,* **37**, 624-630.

Zahn, T.P., Grafman, J., & Tranel, D. (1999). Frontal lobe lesions and electrodermal activity: Effects of significance. *Neuropsychologia*, **37**, 1227-1241.

Zamrini, E.Y., Meador, K.J., Loring, D.W., Nichols, F.T., Lee, G.P., Figueroa, R.E., & Thompson, W.O. (1990). Unilateral cerebral inactivation produces differential left/right heart rate responses. *Neurology,* **40**, 1408-1411.

Zhang, Z.H., Dougherty, P.M., & Oppenheimer, S.M. (1999). Monkey insular cortex neurons respond to baroreceptor and somatosensory convergent inputs. *Neuroscience*, **94**, 351-360.

Zhang, Z.H., & Oppenheimer, P.M. (1998). Insular cortex lesions alter baroreceptor sensitivity in the urethane-anesthetized rat. *Brain Research*, **813**, 73-81.

Zhang, Z.H., & Oppenheimer, P.M. (2000). Electrophysiological evidence for reciprocal insulo-insular connectivity of baroreceptor-related neurons. *Brain Research*, **863**, 25-41.

Zimbardo, P.G. (1988). *Psychology and life* (12th ed.). Glenview, IL: Scott, Foresman & Co.

Zoccolotti, P., Caltagirone, C., Benedetti, N., & Gainotti, G. (1986). Disorders of autonomic responses to emotional stimuli in patients with unilateral hemispherical lesions. *Encephale,* **12**, 263-268.

Zoccolotti, P., Scabini, D., & Violani, C. (1982). Electrodermal responses in patients with unilateral brain damage. *Journal of Clinical Neuropsychology*, **4**, 143-150.

Chapter 9

SEROTONERGIC METABOLISM AND VIOLENCE/AGGRESSION

Gil Zalsman, M.D. and Alan Apter, M.D.
Department of Psychiatry, Tel-Aviv University

INTRODUCTION

The association between a low incidence of serotonin metabolites in the central nervous system and aggressive-suicidal behavior (Asberg, Traskman, & Thoren, 1976) is now well established. Originally, it appeared that serotonin was related to suicidal behavior, but subsequently it was shown that low serotonin turnover is also related to impulsivity and violent behavior (Linnoila & Virkkunen, 1992).

The association of impulsive aggression, alcoholism, personality disorders, suicidal behavior and low 5-hydroxy-indoleacetic acid (5-HIAA) in cerebro-spinal fluid (CSF) has been termed the *low serotonin syndrome* (Linnoila & Virkkunen, 1992). Low 5-HIAA was found to be associated with impulsive aggression in primates (Mehlman *et al.*, 1994) and in human subjects (Kruesi *et al.*, 1990, 1992). An association between serotonin (5-HT) dysfunction and aggression-related behaviors, such as violence, suicidality and impulsivity, has been demonstrated in a number of newer biochemical studies (Arango, Underwood, Gubbi, & Mann, 1995; Apter, van Praag, Sevy, Korn, & Brown, 1990; Coccaro, 1989; Coccaro *et al.*, 1989; Mann, 1998; Mann & Malone, 1997). In adolescents, aggression and violence may be as important in some kinds of suicidal behaviors as is depression (Apter *et al.*, 1995).

Linnoila and Virkkunen (1992) have claimed that impulsive aggression, alcoholism, the circadian rhythm and low serotonin are all mediated by the Suprachiasmatic Nucleus (SCN). This area of the brain is connected to the raphe nuclei through serotonergic pathways, and controls the circadian rhythm (Mann, 1995). Correlations between the circadian rhythm and low serotonin activity in platelets have also been shown in suicidal patients (Verkes, Pijl, Meinders, & van Kempen, 1996).

There has, however, been some controversy surrounding the concept of the low serotonin syndrome (Berman, Tracy, & Coccaro, 1997). For example, Serotonin Selective Reuptake Inhibitors (SSRI) have been disappointing in the prevention of aggression, violence and suicidality (van Praag, 1991). In a comprehensive and critical review of 39 studies linking 5-HIAA and aggression in humans, Balaban, Alper, and Kasamon (1996) concluded that no difference in mean 5-HIAA levels has been found between violent impulsive psychiatric patients and patients diagnosed with other psychiatric and medical conditions, if levels of 5-HIAA were corrected for age, gender and height. However, mean levels of 5-HIAA were significantly lower in healthy volunteers. Balaban *et al.* (1996) concluded that low 5-HIAA levels are connected to psychiatric disorders, but are not specific to impulsive aggression or criminality.

BIOLOGICAL CORRELATES OF OUTWARDLY-DIRECTED AGGRESSION

The study of aggression in both humans and animals has been hampered both by the lack of a set of uniform definitions and by the variation in display of aggressive behavior, depending on the social context in which it is expressed. Moyer (1977) has delineated seven types of aggressive behavior, and has hypothesized that a different neuroanatomical and neuropharmacological substratum underlies each of these subtypes. While such categories are important in understanding aggressive behavior in animals, it has been pointed out that the study of human aggression is complicated by a variety of psychological factors, including unconscious defense mechanisms, and environmental and social factors (van Praag, 1991).

Acts of violence by psychiatric patients are common, and as a result restraint and seclusion procedures as well as medications are used routinely in most hospitals to calm patients. Surveys of seclusion practices indicate that 26% of inpatients are secluded during their stay at an acute service hospital, primarily for assaultive behavior

(Plutchik, Climent, & Ervin, 1976). In addition, there are reports that up to 40% of psychiatrists may have been physically attacked at least once by patients, and some work has recently been done to demonstrate the need for training programs to enhance clinician confidence in coping with possible patient aggression (Kermani, 1981). It seems more and more that violence and homicidal acts are not as rare in psychotic patients as once thought (Feld, Zalsman, & Apter, 2001).

The Serotonin (5-HT) System

In general, data from studies looking at the role of 5-HT in human aggression suggest that aggression and hostility are associated with a decreased level of 5-HT and its metabolite 5-HIAA. The works of Asberg *et al.* (1976) established this observation in human subjects. Similarly, Lidberg, Asberg, and Sundquist-Stensman (1984) reported case histories of 3 individuals who demonstrated aggression towards family members, eventually murdering a child. All 3 had low 5-HIAA levels. Lidberg, Tuck, Asberg, Scalia-Tomba, and Bertilsson (1985) also compared 16 men who had committed homicide, 22 men who had attempted suicide, and 39 controls. CSF 5-HIAA levels were reduced in the suicidal men, especially in those who had used violent means, and in those who had killed a sexual partner (i.e., crimes of passion involving impulsivity and intense affect). Linnoila *et al.* (1983) found that impulsive murderers had lower CSF 5-HIAA levels than did non-impulsive murderers, and multiple murderers had lower CSF 5-HIAA levels than single murderers. Other monoamine metabolites were also decreased in these impulsive murderers, but not significantly so. These studies have also presented evidence suggesting that it may not be the aggressive behavior per se, rather the lack of impulse control that constitutes the behavioral correlate of such deficient 5-HT metabolism.

Linnoila *et al.* (1983) found that 20 arsonists had lower CSF 5-HIAA levels than both habitual offenders and normal controls, while Bioulac, Benezech, Renaud, Noel, and Roche (1980) found that criminals with the 47,XYY syndrome showed a significant correlation, matching high aggression scores to decreased CSF 5-HIAA. Treatment with 5-HT led to clinical improvement. Decreased CSF 5-HIAA has also been demonstrated in mentally retarded individuals who exhibit aggression directed towards others (Greenberg & Coleman, 1976). O'Neil, Page, Adkins, and Eicelman (1986) have reported a case study of a retarded man with Cornelia de Lange syndrome, intense aggressive behavior towards others, and low blood 5-HT levels, for whom treatment with trazodone and tryptophan resulted in an increase of blood 5-HT levels and a decrease of aggressive behavior.

Animal studies have also provided a preponderance of evidence demonstrating that potentiation of 5-HT function decreases aggression in mice and muricidal rats, while decreasing 5-HT function increases aggression in mice (Hodge & Butcher, 1974), thus suggesting the importance of serotonergic mechanisms in the mediation of aggression in these animal models (Hodge & Butcher, 1974; Valzelli, 1981). Other animal studies have provided data suggesting that when a critical balance between the monoaminergic systems is disturbed, animals exhibit specific types of aggressive behavior (Eichelman & Thoa, 1973; Valzelli, 1981), although different investigators have produced variable results when, for instance, producing increased dopamine (DA) synthesis and decreased 5-HT synthesis in mice (Hodge & Butcher, 1975).

It has been suggested that the anti-aggression effect of lithium, shown in both humans and in rats, may be moderated through the 5-HT system, since lithium is known to increase brain uptake of tryptophan, the primary dietary precursor of 5-HT, and is thought to increase brain neurotransmission of 5-HT (de Montigny, Grunberg, Mayer, & Deschenes, 1981). Buspirone was found to lower aggression in inpatient pre-pubertal children, exhibiting moderate to severe aggression (Pfeffer, Jiang, & Domeshek, 1997; Pfeffer, Normandin, & Tatsyuki, 1994).

Finally, clozapine, an atypical anti-psychotic drug that has affinity to both serotonin and dopamine receptors, reduces aggression and violence in violent patients suffering from schizophrenia (Glazer & Dickson, 1998).

Other Neurotransmitter Systems

While a variety of studies investigating the relationship between the catecholamines (CA) and human aggressive behavior have been undertaken, results have often been contradictory (Valzelli, 1981). Some authors found a complex set of changes in biogenic trace amines of violent criminal offenders. Most worthy of note was the decrease in platelet monoamine oxidase (MAO) activity among these individuals. Woodman and Hinton (1978) have reported on a group of sympathetically hyperactive prisoners, who excreted more norepinephrine (NE) and less epinephrine (E) in their urine than did control prisoners and patients. These individuals showed similar differences in plasma CAs. Interestingly, these prisoners had also committed the most violent set of crimes, when compared to the control group of prisoners. Brown, Goodwin, Ballenger, Goyer, and Major (1979) have found a significant positive correlation between CSF 3-methoxy-4-hydroxyphenylglycol (MHPG) and aggressive behavior in military personnel, while Sandler, Ruthven, Goodwin, Field and Matthews (1978) have reported that prisoners with histories of violent crimes had higher levels of urinary phenylacetic acid, than did prisoners with histories of non-violent crimes. More recently, beta-adrenergic antagonists, such as propranolol, have been utilized with some efficacy in a subset of individuals (generally having some type of organic brain injury, or mental retardation) who exhibit ongoing aggressive behaviors (Caspi *et al.*, 2001).

With respect to animal models, Eichelman and Thoa (1973) have assigned aggressive behavior in the rat to 3 categories: predatory, spontaneous, and irritable. Through a series of experiments, they have presented data suggesting that muricidal predatory aggressive behavior is related to both brain 5-HT and to brain acetylcholine levels, spontaneous aggression is related to levels of brain DA or to stimulation of the DA receptors, and irritable aggression is most consistently related to levels of brain NE.

Hodge and Butcher (1975), however, found that giving intra-peritoneal L-dopa in combination with a peripheral decarboxylase inhibitor (thus causing an increase in central DA availability) produced a decrease in isolation-induced aggression in mice. The same results were found with apomorphine and pimozide, while d-amphetamine was found to potentiate aggression at lower dosages, while at higher dosages isolation-induced aggression was decreased. Pucilowski, Kostowski, Bidzinski and Hauptmann (1982) have reported that 6-hydroxydopamine-induced lesions of the ventral mesencephalic tegmental area produce decreased foot-shock-induced fighting, but fail to influence muricidal behavior in rats.

Mice fed L-tyrosine (the metabolic CA precursor) or L-phenylalanine (the amino acid precursor of tyrosine) have been reported to demonstrate increased territorial aggressive behavior (Valzelli, 1981). Drugs that increase central NE activity, such as tricyclic antidepressants (TCAs) and monoamine oxidase inhibitors (MAOIs), have been shown to increase shock-induced fighting in rats, while "sham rage" in cats has been decreased or stopped by drugs that deplete central NE such as reserpine and alpha-methyl-p-tyrosine (Reis, 1972). Finally, Ciaranello, Lipsky, and Axelrod (1974) have presented data suggesting that, in specific strains of mice, there may be genetic differences that affect both fighting behavior and levels of tyrosine hydroxylase, dopamine-beta-hydroxylase, and phenylethanolamine N-methyltransferase, all of which are involved in the biosynthesis of the catecholamines.

Animal studies have provided increasing evidence indicating that the GABA/benzodiazepine system may function to modulate aggressive behavior. Increasing central GABAergic activity appears to decrease some types of aggressive behavior in rats and mice, while GABA antagonists appear to induce aggressive behavior. Aggressive mice may also have lower regional GABA concentrations than their non-aggressive counterparts. However, it should be noted that one group of researchers has presented data suggesting that while higher doses of GABA-potentiating compounds or GABA-agonists suppress mouse-killing behavior in rats, smaller doses may facilitate this behavior in some animals (DePaulis & Vergnes, 1983, 1984). They suggest that this may be the result of the anticonflict effects of increasing GABA levels, rather than from an active enhancement of muricidal behavior. In addition, Potegal, Yoburn, and Glusman (1983) have shown that intracranial administration of GABA-modulating agents produces differing effects upon muricide, depending upon the specific neuroanatomical region affected. They suggest that, since GABA is an inhibitory neurotransmitter, GABAergic activation in a nuclear region excitatory to aggression will be inhibitory, while GABAergic activation in a nuclear region inhibitory to aggression will be disinhibitory—which seems to occur in some areas of the septum and the caudal ventral tegmental area.

Benzodiazepines—drugs which stimulate the GABA/benzodiazepine receptor and thus increase central GABAergic function—have been shown to reduce aggressive behavior in humans and in a wide variety of animals, although there is also evidence that chronic low dosage administration of these drugs can cause increased aggressive behavior in humans and animals (DePaulis & Vergnes, 1984; Rudissaar, Pruus, Skrebuhhova-Malmros, Allikmets, & Matto, 2000; Ryan, 2000).

More recently, there is also evidence that other neurotransmitter systems may play some role in aggression. Acetylcholine has been implicated, with some data suggesting that cholinomimetic drugs such as carbachol increase aggression when injected into the amygdala of cats; physostigmine, a cholinesterase inhibitor, increases aggression when injected into the amygdala of rats; and cholinergic agonists induce aggression when injected into the lateral hypothalamus of rats, while this effect is blocked with atropine (Ryan, 2000).

BIOLOGICAL LINKS BETWEEN AGGRESSION AND SUICIDALITY

The importance of suicidology for psychiatry needs no elaboration. However, the "epidemic" of suicidal behavior in the young in recent years (Shaffer, 1988) has made this a major public health issue as well. Suicide, homicide, and car accidents are now the leading causes of death in the 15-24 year old age group (Pfeffer, 1986). Prevention programs have proved ineffective and may even be dangerous, as they have been reported to increase suicide, probably via a "Werther" effect (Gould, Shaffer & Prudence, 1992).

The relationship between suicide and aggression has long been recognized by psychoanalysts, and has been best formulated by Menninger (1933), who proposed that a dynamic triad underlies all aggressive behavior, whether directed inwards or outwards: the wish to die, the wish to kill, and the wish to be killed. These are all seen to be derivatives of Freud's concept of "thanatos," or the death instinct.

There are many reports in the daily press of murderers who commit suicide. Roughly one out of four patients with a history of violent actions in the past has made a suicide attempt (Bennett, 2000). According to various authors, from 7% to 48% of those patients with a history of violent behavior have also made suicide attempts in the past (Reza, Mercy, & Krug, 2001). This is reported to be true for adults and for prepubertal children (Pfeffer, 1986; Plutchik, van Praag, Conte, & Picard, 1989). Similar findings have been reported in prisoners (Plutchik *et al.*, 1976).

Looking at the issue in terms of patients who are predominantly suicidal, Tardiff and Sweillam (1980) found that of a large group of suicidal patients admitted to mental hospitals, 14% of males and 7% of females were assaultive at the time of or just prior to admission. Similarly, on the basis of detailed studies of psychotic patients, Kermani (1981) has identified what he calls the "violent depressive type," who has a long history of suicide attempts as well as a history of violence. It has likewise been reported that female suicide attempters show more hostility and engage in more arguments and friction with friends and relatives than do a comparable group of non-suicidal depressed women. Hospitalized depressed patients with a history of self-destructive acts have been found to have high levels of hostility and violence, as measured by their need for seclusion and restraint (Weissman, Fox, & Klerman, 1973).

In a large group of psychiatric inpatients with mixed diagnoses, about 40% had made a suicide attempt, 42% had engaged in violent behavior in the past, and 23% had histories of both types of behavior (Plutchik, van Praag, & Conte, 1985). Almost every one of the 30 variables measured in these patients turned out to be a significant predictor of both suicide risk and violence risk. These results strongly support the idea of a close link between suicide risk and violence risk, regardless of diagnosis.

Another important finding in this regard is that of Shaffer (1988), who found that a combination of depressive symptoms and antisocial behavior was the most common antecedent of teenage suicide.

The Serotonin (5-HT) System

Asberg *et al.* (1976) reported on 44 patients with endogenous depression and 24 patients with reactive depression, demonstrating a high correlation linking decreased CSF 5-HIAA—the primary metabolite of 5-HT—and history of suicide attempts, particularly those that involved a violent component, during the illness. This finding has been confirmed by a number of other researchers (Banki & Arato, 1983; Oreland *et al.*, 1981). Agren (1980) has also reported that low CSF 5-HIAA correlated not only with suicidal ideation, but also with overt anger, and suggested that this might provide evidence in support of the hypothesis that depressives with low CSF 5-HIAA are more prone to violent suicidal acts than depressives without low CSF 5-HIAA.

Depressives with low CSF 5-HIAA appear to be not only more suicidal but also more aggressive (van Praag, Lemus, & Kahn, 1986). Two groups of melancholic patients were compared on several aggression measures, as well as on CSF 5-HIAA levels. Depressives were divided into two groups, one with normal and the other with low post-probenecid CSF 5-HIAA levels. The latter group had made more suicide attempts and had demonstrated higher hostility ratings on several measures than had the former group. Mann *et al.* (1992) have demonstrated a relationship between central and peripheral serotonin and fenfluramine-stimulated maximal prolactin response, in depressed and suicidal patients.

Interestingly, van Praag (1983), Ninan *et al.* (1984) and Banki and Arato (1983) have all reported significantly lower concentrations of CSF 5-HIAA in suicidal schizophrenics, with a correlation linking recent suicide attempts and lower 5-HIAA, thus suggesting that the psychopathological correlate of decreased CSF 5-HIAA might involve suicidality across psychiatric diagnoses. However, it should be noted that others could not replicate these findings, nor could Pickar *et al.* (1986), although these latter investigators did show a relationship between lower CSF 5-HIAA and higher drug-free hostility and uncooperativeness ratings in these schizophrenics. In addition, Brown and colleagues (Brown, Ebert *et al.*, 1982; Brown, Goodwin, & Bunney, 1982) reported that a history of aggressive behavior, a history of suicide attempts, and low CSF 5-HIAA were all positively correlated among patients with personality disorders, while Rydin, Schalling, and Asberg (1982) showed that depressed and/or suicidal patients with low CSF 5-HIAA had higher hostility scores and a lower anxiety tolerance, as determined by Rorschach responses, than did their counterparts with normal CSF 5-HIAA values. Mann (1998) reviewed the current data on the psychobiology of suicidal behavior, and once again put the serotonin system at the center of interest for future research.

The Catecholamine (CA) Systems

Brown *et al.* (1979) reported increased levels of 3-methoxy-4-hydroxyphenylglycol, or MHPG, a norepinephrine (NE) metabolite, in patients with personality disorders and life histories of aggressive and suicidal behavior, suggesting the involvement of the NE system in aggression regulation. However this finding has not been replicated.

Ostroff *et al.* (1982) studied the 24-hour urinary norepinephrine/epinephrine (NE/E) ratio in psychiatric patients, and reported interesting peripheral data on the NE system and aggression. Their work was based on an expanded model, indicating differences in psychophysiological responses (e.g., electrodermal activity) in subjects who reported anger against others in contrast to subjects who reported anger against themselves during stressful situations. Ostroff *et al.* (1982) hypothesized that "anger out" was associated with NE-like substances and "anger in" was associated with E-like substances. Three patients who made violent suicide attempts were shown to have a significantly lower NE/E ratio than did other patients, thus appearing to confirm this hypothesis.

NEW RESEARCH DATA

Recent advances in molecular biology such as the discovery of serotonin-related polymorphisms have opened up an exciting new field of research.

Serotonin Transporter Gene Polymorphism

One candidate gene potentially influencing aggressive suicidal behaviors is the Serotonin Transporter (5-HTT) gene (Furlong, Ho, Walsh *et al.*, 1988). This gene codes for the serotonin transporter which is localized in the outer membrane of the pre-synaptic serotonergic neurons (Heils, Mossner, & Lesch, 1997; Ramamoorthy *et al.*, 1993).

The serotonin transporter is a sodium-dependent transporter with high affinity for serotonin. It has an important role in controlling serotonin availability in the synapse by regulating the re-uptake of serotonin. This protein is the major target of serotonin re-uptake inhibitor anti-depressant medications (Heils *et al.*. 1997). Some drugs are aggregated in the serotonergic neurons by the 5-HTT (Lesch, Wolozin, Estler, Murphy, & Riederer, 1993).

The gene encoding the 5-HTT protein gene (SLC6A4) was mapped onto the proximal long arm of chromosome 17 close to 17q12 (Gelernter, Pakstis, & Kidd, 1995; Lesch *et al.*, 1993, 1994; Ramamoorthy *et al.*, 1993). Lesch *et al.* (1994) have shown that the 5-HTT gene has 14 exons and 12 introns. In the second intron, there is a region of variable number tandem repeats (VNTR) of 17 base pairs (bp) in the second intron with three major alleles. Another polymorphism found in the promoter region is the 5-HTT linked polymorphic region (5-HTTLPR) which consists of two common alleles, a short (S) and a long (L) variant, differing by repeats of 20-23GC44 base pairs (Heils *et al.*, 1996).

Lesch *et al.* (1995) have shown that there are no differences in the amino acid sequence order between mood disorder, affectively ill probands and a healthy population. No major mutation was found in the 5-HTTSLC6A4 coding area region between probands and control groups subjects (Di Bella, Catalano, Balling, Smeraldi, & Lesch, 1996). Since no constructive structural amino acid sequence changes were found in the gene encoding the 5-HTT protein, it was hypothesized that a dysfunction or alteration in the expression of the gene may be found in serotonin-related psychopathologies.

Alcohol dependency and ethanol tolerance were also found to be associated with the S allele in a few studies (Lesch & Mossner, 1998; Turker *et al.*, 1998), but showed no association in other studies (Gelernter, Kranzler, & Cubells, 1997). Hallikainen *et al.* (1999) found an association between the S allele and type 2 alcoholism with habitual impulsive violent behavior. Some evidence has been found in connection with Alzheimer disease (Lesch & Mossner, 1998).

The role of the 5-HTTLPR polymorphism in suicidal behavior was studied by Russ, Lachman, Kashdan, Saito, and Bajmakovic-Kacila (2000) and by Geijer *et al.* (2000). In both case-control studies, there was no significant difference in the frequency of the 5-HTTLPR5-HTT promotor alleles between Caucasian European patients who attempted suicide and a control group. However, Du *et al.* (1999) reported a modest association between the L allele of the 5-HTTLPR and suicidality in depressed patients, while Bondy, Erfuth, Jonge, Kruger, & Meyer (2000) found a possible association of the S allele with violent suicide victims.

Zalsman, Frisch, Bromberg *et al.* (2001) examined the association of 5-HTTLPR with suicidal behavior and related traits in Israeli suicidal adolescent inpatients, using the haplotype relative risk (HRR) method that controls for artifacts caused by population stratification. In this method, parents are used as controls for the proband studied. Forty-eight inpatient adolescents who recently attempted suicide were assessed by structured interviews for detailed clinical history, diagnoses, suicide intent, suicide risk, impulsivity, violence and depression. Blood samples were collected and DNA was extracted from patients and their biological parents. The 5-HTTLPR allele frequencies were tested for association with suicidality by the HRR method. In addition, the relationship between genotypes and phenotypic severity of several clinical parameters was analyzed. No significant allelic association of the 5-HTTLPR polymorphism with suicidal behavior was found (Chi square = 0.023, $p = .88$). Analysis of variance of the suicide-related trait measures for the three genotypes demonstrated a significant difference in violence measures between patients carrying the LL and LS genotypes (9.50 ± 4.04 [SD] vs. 5.36 ± 4.03; $p = .029$). This study suggests that the 5-HTTLPR polymorphism is unlikely to have major relevance for the pathogenesis of suicidal behavior in adolescence, but may contribute to violent behavior in this population. Such an association was found for violence traits as measured by the Past Feelings and Acts of Violence Scale (PFAVS). This questionnaire has been found to be highly sensitive and reliable in detecting aggression and violence (Plutchik & van Praag, 1990)—two important factors in the evaluation of suicidal adolescents (Apter, Plutchik, & van Praag, 1993; Apter *et al.*, 1995; Coccaro, 1989; Coccaro *et al.*, 1989; Gould, Shaffer, & Prudence, 1992). The PAFVS significantly discriminated between psychiatric inpatients with and without violent behavior (Plutchik & van Praag, 1986). In a previous study, a subgroup of violent patients showed higher suicide risk and impulsivity compared with the non-violent subgroup (Apter *et al.*, 1995). However, it should be noted that two studies (Coccaro *et al.*, 1989) emphasized that impulsive aggressive behavior is an essential component of suicidality, and is responsible for the serotonin dysregulation. In this study, an association was found between the serotonergic genotype and aggressive behavior, but not with suicidality.

If the LL genotype (but not the LS and SS genotypes) is significantly more frequent in the high-risk group of aggressive suicidal adolescents, then the mechanism of inheritance of suicidality is apparently more complex than a Mendelian recessive or co-dominance model.

The Tryptophan Hydroxilase Gene Polymorphism

Based on genetic studies and the findings relating serotonin to suicidal behavior, some researchers have suggested that the gene coding for the rate-limiting enzyme in serotonin metabolism, tryptophan hydroxylase (TPH), is a candidate gene for suicidal behavior.

Nielsen, Dean, and Goldman (1992) placed the human TPH gene in the linkage map on chromosome 11. Using the single strand conformational polymorphism (SSCP) technique, they identified a polymorphism in intron 7, and designated the alleles U and L, depending on their SSCP migration. In a study of impulsive violent alcoholics, the same group (Nielsen *et al.*, 1994) reported an association between the TPH genotype and CSF levels of 5-HIAA. The polymorphism was found to be associated with a history of suicide attempts and the number of suicide attempts, independent of psychiatric diagnosis, impulsivity, or 5-HIAA levels. The L polymorphic allele was significantly associated with suicidality. Abbar *et al.* (1995) criticized these findings, suggesting that these data are not sufficient to support the existence of an association between the TPH gene and suicidal behavior.

Different results were found by Mann *et al.* (1997), who reported a possible association of the U allele but not the L allele, in 52 American Caucasian non-impulsive major depressed inpatients, with and without a history of suicide attempts. Several years later, however, Nielsen, Jenkins, Stefasko, Jefferson, and Goldman (1997) and Nielsen, *et al.* (1998) replicated their own results in another group of Finish patients, using polymerase chain reaction-restriction fragment length polymorphism (PCR-RFLP) assays for, essentially, the same polymorphic system. They demonstrated that the SSCP polymorphism

(U/L), which they had identified previously, corresponds to two single nucleotide polymorphisms (SNPs), A218C and A779C in intron 7. Studies used a PCR-RFLP method to detect what was originally described as an SSCP polymorphism to actually detect A218C or A779C (Nielsen *et al.*, 1997). TPH 779A and 218A, correspond to the U TPH allele, and the TPH 779C and 218C correspond to the L TPH allele (Nielsen *et al.*, 1992). Paralleling their earlier findings, Nielsen *et al.* (1998) found that the TPH 779C allele was associated with suicidality. This study also used a sib-pair analysis (369 sib pairs) which yielded a significant linkage to suicidal behavior and severe suicide attempts.

Several other groups have reported results that were mixed or negative. Also using RFLP, a Swiss-French group (Bellivier *et al.*, 1998) recently reported an association between the TPH gene and the bipolar disorder in intron 7 polymorphism, but no association between A218C polymorphism and suicidal behavior. Later, they published a re-analysis of their data in an extension of the original sample, suggesting that the association they reported between bipolar affective disorder and the TPH gene A218C polymorphism, was due to a subgroup of patients with a history of suicidal behavior. In this study, the A allele frequency was even higher in the subgroup of patients who had made a violent suicide attempt (Bellivier *et al.*, 1998). Thereafter, the same group performed a case-control study of 236 suicide attempters matched with 161 controls, and found an association to the allele A of the A218C polymorphism (Buresi, Courtet, Leboyer, Feingold, & Malafosse, 1997).

Furlong, Ho, Rubinsztein *et al.* (1998) failed to associate the TPH gene with suicide in patients with affective disorder. Manuck *et al.* (1999) reported a significantly higher score for the TPH U allele on measures of aggression and anger, traits that are thought to be connected to suicidal behavior, and Kunugi *et al.* (1999), in a Japanese population, reported no evidence for an association between TPH A218C and A779C polymorphisms and suicidal behavior in 46 suicide attempters and 208 healthy controls. In a case-control association study in a French population, Abbar *et al.* (1995) compared 62 suicide attempters with 52 healthy controls and noted no association between TPH polymorphism and suicide attempts, but since they studied a different polymorphism (an RFLP recognized by Ava II), their results cannot be compared directly to those of Nielsen or any of the other studies discussed here (Nielsen, 1996). Zalsman, Frisch, King *et al.* (2001) failed to demonstrate the association of TPH polymorphism A218C to suicidality, aggression or impulsivity in the Jewish population. Nolan, Volavka, Lachman, and Saito (2000) found an association of TPH L allele with violence in schizophrenic males.

CONCLUSIONS

The Aspen neurobehavioral conference consensus statement that was concluded during the spring of 1998 and 1999 was published in 2001 (Filley *et al.*, 2001). The statement declared that violence can result from brain dysfunction, especially frontal lobe dysfunction, altered serotonin metabolism and heredity. The role of aggression, both inwardly and outwardly directed, in the various major psychiatric disorders has not yet been addressed in a systematic enough manner. Thus, both because of obvious moral, legal, and sociological issues, and because of its prevalence among psychiatric patients, the study of aggression is emerging as an increasingly important area of psychiatric research. This chapter has indicated some of the directions which biological research may take in order to implement the Aspen declaration. It is our expectation that advances will be made in a number of different areas, including biochemistry, genetics, neuroimaging and pharmacology.

REFERENCES

Abbar, M., Courted, P., Amadeo, S., Caer, Y., Mallet, J., Baldy-Moulinier, M., Castlenau D., & Malefosse, A. (1995). Suicidal behaviors and the tryptophan hydroxylase gene. *Archives of General Psychiatry*, **52**, 846-849.

Agren, H. (1980). Symptom patterns in unipolar and bipolar depression correlating with monoamine metabolites in the cerebrospinal fluid: I. General patterns. *Psychiatry Research*, **3**, 211-223.

Apter, A., Gothelf, D., Orbach, I., Har-Even, D., Weizman, R., & Tyano, S. (1995). Correlation of suicidal and violent behavior in different diagnostic categories in hospitalized adolescent patients. *Journal of the American Academy for Child and Adolescent Psychiatry*, **34**, 7, 912-918.

Apter, A., Plutchik, R., & van Praag, H.M. (1993). Anxiety, impulsivity and depressed mood in relation to suicide and violent behavior. *Acta Psychiatrica Scandinavica*, **87**, 1-5.

Apter, A., van Praag, H.M., Sevy, S., Korn, M., & Brown, S. (1990). Interrelationships among anxiety, aggression, impulsiveness and mood: A serotonergically linked cluster? *Psychiatric Research*, **32**, 191-199..

Arango, V., Underwood, M.D., Gubbi, A.V., & Mann, J.J.(1995). Localized alterations in pre- and postsynaptic serotonin binding sites in the ventrolateral prefrontal cortex of suicide victims. *Brain Research*, **688**, 121-133.

Asberg, M., Traskman, L., & Thoren, P. (1976). 5-HIAA in the cerebrospinal fluid: A biochemical suicide predictor? *Archives of General Psychiatry*, **33**, 1193-1197.

Balaban, E., Alper, J.S., & Kasamon, Y.L. (1996). Mean genes and the biology of aggression: a critical review of recent animal and human research. *Journal of Neurogenetics*, **11**, 1-43.

Banki, C.M., & Arato, M. (1983). Amine metabolites, neuroendocrine findings and personality dimensions as correlates of suicidal behavior. *Psychiatric Research*, **10**, 253-261.

Bellivier, F., Leboyer, M., Courtet, P., Buresi, C., Beaufils, B., Samolyk, D., & Allilaire, J.F. (1998). Association between the tryptophan hydroxylase gene and manic-depressive illness. *Archives of General Psychiatry*, **55**, 33-37.

Bennett, J.A. (2000). Impact of the Brady Act on homicide and suicide rates. *Journal of the American Medical Association*, **284**, 2720-2721.

Berman, M.E., Tracy, J.I., & Coccaro, E.F. (1997). The serotonin hypothesis of aggression revisited. *Clinical Psychology Review*, **17**, 651-665.

Bioulac, B., Benezech, M., Renaud, B., Noel, B., & Roche, D. (1980). Serotonergic functions in the 47, XYY syndrome. *Biological Psychiatry*, **15**, 917-923.

Bondy, B., Erfuth, A., Jonge, S., Kruger, M., & Meyer, H. (2000). Possible association of the short allele of the serotonin transporter promoter gene polymorphism (5-HTTLPR) with violent suicide. *Molecular Psychiatry*, **5**, 193-195.

Brown, G.L., Ebert, M.H., Goyer, P.F., Jimerson, D.C., Klein, W.J., Bunney, W.E., & Goodwin, F.K. (1982). Aggression, suicide and serotonin: Relationships to CSF amine metabolites. *American Journal of Psychiatry*, **139**, 741-746.

Brown, G.L., Goodwin, F.K., Ballenger, J.C., Goyer, P.F., & Major, L.F. (1979). Aggression in humans correlates with cerebrospinal fluid amine metabolites. *Psychiatry Research*, **1**, 131-139.

Brown, G.L., Goodwin, F.K., & Bunney, W.E. (1982). Human aggression and suicide: Their relationship to neuropsychiatric diagnoses and serotonin metabolism. In B.T. Ho, J.C. Schoolar, & E. Usdin (Eds.), *Serotonin in biological psychiatry: Advances in biological psychiatry*. New York: Raven Press.

Buresi, C., Courtet, P., Leboyer, M., Feingold, J., & Malafosse, A. (1997). Association between suicide attempt and the tryptophane hydroxylase Gene. *American Journal of Human Genetics*, **61**(Supp.), 270.

Caspi, N., Modai, I., Barak, P., Waisbourd, A., Zbarsky, H., Hirschmann, S., & Ritsner, M. (2001). Pindolol augmentation in aggressive schizophrenic patients: a double-blind crossover randomized study. *International Journal of Clinical Psychopharmacology*, **16**, 111-115.

Ciaranello, R.D., Lipsky, A., & Axelrod, J. (1974). Association between fighting behavior enzyme activity in two inbred mouse sublines. *Proceedings of the National Academy of Sciences of the USA*, **71**, 3006-3008.

Coccaro, E.F. (1989). Central serotonin and impulsive aggression. *British Journal of Psychiatry*, **155**, 52-62.

Coccaro, E.F., Siever, L., Howard, M., Klar, H., Maurer, G., Cochrane, K., Cooper, T.B., Mohs, R.C., & Davis, K.L. (1989). Serotonergic studies in patients with affective and personality disorders. *Archives of General Psychiatry*, **46**, 587-599.

de Montigny, C., Grunberg, F., Mayer, A., & Deschenes, J.P. (1981). Lithium induces rapid relief of depression in tricyclic antidepressant drug non-responders. *British Journal of Psychiatry*, **138**, 252-256.

DePaulis, A., & Vergnes, M. (1983). Induction of mouse-killing in a rat by intrava ventricular injection of GABA agonist. *Physiology and Behavior*, **30**, 383-388.

DePaulis, A., & Vergnes, M. (1984). GABAergic modulation of mouse-killing in a rat. *Psychopharmacology*, **83**, 367-372.

Di Bella, D., Catalano, M., Balling, U., Smeraldi, E., & Lesch, K.P. (1996). Systematic screening for mutations in the coding region of the human serotonin transporter (5-HTT) gene using PCR and DGGE. *American Journal of Medical Genetics*, **67**, 541-545.

Du, I. , Faludi, G., Palkovitz, M., Demeter, E., Bakish Dlapierre, Y.D., Lapierre, Y.D, *et al.* (1999). Frequency of long allele in serotonin transporter gene is increased in depressed suicide victims. *Biological Psychiatry*, **46**,196-201.

Eichelman, B.S.,Jr, & Thoa, N.B. (1973). The aggressive monoamines. *Biological Psychiatry*, **6**, 143-164.

Feld, J., Zalsman, G., & Apter, A. (2001). Violence and homicide in psychotic patients: is this a variable phenomenon? *Harefua* [Hebrew], **140**, 51-54.

Filley, C.M., Price, B.H., Nell, V., Antoinette, T., Morgan, A.S., Bresnahan J.F., Pincus, J.H., Gelbort, M.M., Weissberg, M., & Kelly, J.P. (2001). Toward an understanding of violence: neurobehavioral aspects of unwarranted physical aggression: Aspen neurobehavioral conference consensus statement. *Neuropsychiatry, Neuropsychology & Behavioral Neurology*, **14**, 1-14.

Furlong, R.A., Ho, L., Rubinsztein, J.S., Walsh, C., Paykel, E.S., & Rubinsztein, D.C.(1998). No association of the tryptophan hydroxylase gene with bipolar affective disorder, unipolar affective disorder, or suicidal behavior in major affective disorder. *American Journal of Medical Genetics*, **81**, 245-247.

Furlong, R.A., Ho, L., Walsh, C., Rubinsztein, J.S., Jain, S., Paykel, E.S., Rubinsztein, D.C., *et al.*(1998). Analysis and meta-analysis of two serotonin transporter gene polymorphisms in bipolar and unipolar affective disorders. *American Journal of Medical Genetics*, **81**, 58-63.

Geijer, T., Frisch, A., Persson, M.L., Wasserman, D., Rockah, R., Michaelovsky, E., Apter, A., Jonsson, E., Nothen, M.M., & Weizman, A. (2000). Search for association between suicide attempt and serotonergic polymorphisms. *Psychiatry Genetics*, **10**, 19-26.

Gelernter, J., Pakstis, A.J., & Kidd, K.K. (1995). Linkage mapping of serotonin transporter protein gene SLC6A4 on chromosome 17. *Human Genetics*, **95**, 677-680.

Gelernter, J., Kranzler, H., & Cubells, J.F.(1997). Serotonin transporter protein gene allele and haplotype frequencies and linkage disequilibria in African- and European-American and Japanese populations and in alcohol dependent subjects. *Human Genetics*, **101**, 243-246.

Glazer, W.M., & Dickson, R.A. (1998). Clozapine reduces violence and persistent aggression in schizophrenia. *Journal of Clinical Psychiatry*, **3**, 8-14.

Gould, M.S., Shaffer, D., & Prudence, F. (1992). The clinical prediction of adolescent suicide. In R.W. Maris (Ed.), *Assessment and prediction of suicide*. New York: Guilford Publications.

Greenberg, A.S., & Coleman, M. (1976). Depressed 5-hydroxyindole levels associated with hyperactive and aggressive behavior. Relationship to drug response. *Archives of General Psychiatry*, **33**, 331-336.

Hallikainen, T., Saito, T., Lachman, H.M., Volavka, J., Pohjalainen, T., Ryynanen, O.P., Kauhanen, J., Syvalahti, E., Hietala, J., & Tiihonen, J. (1999). Association between low activity serotonin transporter genotype and early onset alcoholism with habitual impulsive violent behavior. *Molecular Psychiatry*, **4**, 385-388.

Heils, A., Mossner, R., & Lesch, K.P. (1997). The human serotonin transporter gene polymorphism—basic research and clinical implications. *Journal of Neural Transmission*, **104**, 1005-1014.

Heils, A., Teufel, A., Petri, S., Stober, G., Riederer, P., Bengel, D., & Lesch, K.P. (1996). Allelic variation of human serotonin transporter gene expression. *Journal of Neurochemistry*, **66**, 2621-2624.

Hodge, G.K., & Butcher, L.L. (1974). 5-Hydrytryptamine correlates of isolation-induced aggression in mice. *European Journal of Pharmacology*, **28**, 326-337.

Hodge, G.K., & Butcher, L.L. (1975). Catecholamine correlates of isolation-induced aggression in mice. *European Journal of Pharmacology*, **31**, 81-93.

Kermani, E.J. (1981). Violent psychiatric patients: a study. *American Journal of Psychotherapy*, **35**, 215-225.

Kruesi, M.J., Hibbs, E.D., Zahn, T.P., Keysor, C.S., Hanburger, S.D., Bartko, J.J., & Rapoport, J.L. (1992). A 2-year prospective follow up of children and adolescents with disruptive behavior disorders. Prediction by CSF 5HIAA, HVA and autonomic measures? *Archives of General Psychiatry*, **49**, 429-435.

Kruesi, M.J., Rapoport, J.L., Hamburger, S., Hibbs, E., Potter, W.Z., Lenane, M., & Brown, G.I. (1990). Cerebrospinal fluid monoamine metabolites, aggression and impulsivity in disruptive behavior disorders of children and adolescents. *Archives of General Psychiatry*, **47**, 419-426.

Kunugi, H., Ishida, S., Kato, T., Tatsumi, M., Hirose, T., & Nank, S. (1999). No evidence for association of polymorphisms of the tryptophan hydroxlase gene with affective disorders or attempted suicide among Japanese patients. *American Journal of Psychiatry*, **156**, 774-776.

Lesch, K.P., Wolozin, B.L., Estler, H.C., Murphy, D.L., & Riederer, P. (1993). Isolation of a cDNA encoding the human brain serotonin transporter. *Journal of Neural Transmission* [Gen Sect], **91**, 67-72.

Lesch, K.P., Balling, U., Gross, J., Strauss, K., Wolozin, B.L., Murphy, D.L., & Riederer, P. (1994). Organization of the human serotonin transporter gene. *Journal of Neural Transmission* [Gen Sect], **95**, 157-162.

Lesch, K.P., Gross, J., Franzek, E., Wolozin, B.L., Riederer, P., & Murphy, D.L. (1995). Primary structure of the serotonin transporter in unipolar depression and bipolar disorder. *Biological Psychiatry*, **37**, 215-223.

Lesch, K.P., & Mossner, R. (1998). Genetically driven variation in serotonin uptake: is there a link to affective spectrum, neurodevelopmental, and neurodegenerative disorders? *Biological Psychiatry*, **44**, 179-192.

Lidberg, L., Asberg, M., & Sundquist-Stensman, U.B. (1984). 5-Hydroxyindoleacetic acid in attempted suicides who have killed their children. *Lancet*, **2**, 928.

Lidberg, L., Tuck, J.R., Asberg, M., Scalia-Tomba, G.P., & Bertilsson, L. (1985). Homicide, suicide and CSF 5-HIAA. *Acta Psychiatrica Scandinavica*, **71**, 230-236.

Linnoila, M., Virkkunen, M., Scheinin, M., Nuutila, A., Rimon, R., & Goodwin, F.K. (1983). Low cerebrospinal fluid 5-hydroxyindoleacetic acid concentration differentiates impulsive from nonimpulsive violent behavior. *Life Sciences*, **33**, 2609-2614.

Linnoila, V.M., & Virkkunen, M. (1992). Aggression suicidality, and serotonin. *Journal of Clinical Psychiatry*, **53**, 46-51.

Mann, J.J. (1995). Violence and aggression. In E.F. Bloom & D.J. Kupfer (Eds.), *Psychopharmacology: the fourth generation of progress* (pp. 1919-1928). New York: Raven Press.

Mann, J.J. (1998). The neurobiology of suicide. *Nature Medicine*, **4**, 25-30.

Mann, J.J.,& Malone, K.M.(1997). Cerebrospinal fluid amines and higher lethality suicide attempts in depressed inpatients. *Biological Psychiatry*, **41**, 162-171.

Mann, J.J., Malone, K.M., Nielsen, D.A., Goldman, D., Erdos, J., & Gelernter, J. (1997). Possible association of a polymorphism of the tryptophan hydroxylase gene with suicidal behavior in depressed patients. *American Journal of Psychiatry*, **154**, 1451-1453.

Mann, J.J., McBride, P.A., Brown, R.P., Linnoila, M., Leon, A., De Meo, M., Mieczkowski, T. Myers, J., & Stanley, M. (1992). Relationship between central and peripheral serotonin indexes in depressed and suicidal psychiatric inpatients. *Archives of General Psychiatry*, **49**, 442-446.

Manuck, S.B., Flory, J.D., Ferrell, R.E., Dent, K.M., Mann, J.J., & Muldoon, M.F. (1999). Aggression and anger-related traits associated with a polymorphism of the tryptophan hydroxylase gene. *Biological Psychiatry*, **45**, 603-614.

Mehlman, P.T., Higley, J.D., Faucher, I., Lilly, A.A., Taub, D.M., Vickers, J., Suomi, S.J., & Linnoila, M. (1994). Low CSF 5-HIAA concentrations and severe aggression and impaired impulse control in nonhuman primates. *American Journal of Psychiatry*, **151**, 1485-1491.

Menninger, K. (1933). *Man against himself.* New York: Harcourt Brace.

Moyer, K.E. (1977). A model of aggression with implication of research. *Psychopharmacology Bulletin*, **13**, 14-15.

Nielsen, D.A. (1996). TPH replication study: not!. *Archives of General Psychiatry*, **53**, 964-965.

Nielsen, D.A., Dean, M., & Goldman, D. (1992). Genetic mapping of the human tryptophan hydroxylase gene on chromosome 11, using an intronic conformational polymorphism. *American Journal of Human Genetics*, **51**, 1366-1371.

Nielsen, D.A., Goldman, D., Virkkunen, M., Tokola, R., Rawlings, R., & Linnoila, M. (1994). Suicidality and 5-hydroxyindoleacetic acid concentration associated with a tryptophan hydroxylase polymorphism. *Archives of General Psychiatry*, **51**, 34-38.

Nielsen, D.A., Jenkins, G.L., Stefasko, K.M., Jefferson, K.K., & Goldman, D. (1997). Sequence, splice site and population frequency distribution analyses of the tryptophan hydroxylase intron 7. *Molecular Brain Research*, **45**, 145-148.

Nielsen, D.A., Virkkunen, M., Lappalainen, J., Eggert, M., Brown, G.L., Long, J.C., Goldman, D., & Linnoila, M. (1998). Tryptophan hydroxylase gene marker for suicidality and alcoholism. *Archives of General Psychiatry*, **55**, 593-602.

Ninan, P.T., van Kammen, D.P., Scheinin, M., Linnoila, M., Bunney, W.E., & Goodwin, F.K. (1984). CSF 5-hydroxyindoleacetic acid levels in suicidal schizophrenic patients. *American Journal of Psychiatry*, **141**, 566-569.

Nolan, K.A., Volavka, J., Lachman, H.M., & Saito, T. (2000). An association between a polymorphism of the tryptophan hydroxylase gene and aggression in schizophrenia and schizoaffective disorder. *Psychiatric Genetics*, **10**, 109-115.

O'Neil, M., Page, N., Adkins, W.N., & Eicelman, B. (1986). Tryptophan-trazodone treatment of aggressive behavior. *Lancet*, **11**, 859-860.

Oreland, L., Wiberg, A., Asberg, M., Traskman, L., Sjostrand, L., Thoren, P., Bertilsson, L., & Tybring, G. (1981). Platelet MAO activity and monoamine metabolites in cerebrospinal fluid in depressed and suicidal patients and in healthy controls. *Psychiatry Research*, **4**, 21-29.

Ostroff, R., Giller, E., Bonese, K., Ebersole, E., Harkness, L., & Mason., J. (1982). Neuroendocrine risk factors of suicidal behavior. *American Journal of Psychiatry*, **139**, 1323-1325.

Pfeffer, C.R. (1986). *The suicidal child*. New York: Basic Books.

Pfeffer, C.R., Normandin, L., & Tatsyuki, K. (1994). Suicidal children grow up: suicidal behavior and psychiatric disorders among relatives. *Journal of the American Academy of Child and Adolescent Psychiatry*, **33**, 1087-1097.

Pfeffer, C.R., Jiang, H., & Domeshek, L.J. (1997). Buspirone treatment of psychiatrically hospitalized prepubertal children with symptoms of anxiety and moderately severe aggression. *Journal of Child and Adolescent Psychopharmacology*, **7**, 145-155.

Pickar, D., Roy, A., Breier, A., Doran, A., Wolkowitz, O., Colison, J., & Agren, H. (1986). Suicide and aggression in schizophrenia. In J.J. Mann & M. Stanley (Eds.), *Psychobiology of suicidal behavior* (pp. 189-196). New York: Annals of the New York Academy of Sciences.

Plutchik, R., Climent, C., & Ervin, F. (1976). Research strategies for the study of human violence. In W.L. Smith & A. Kling (Eds.), *Issues in brain/behavior control*. New York: Spectrum Books.

Plutchik, R., & van Praag, H.M. (1986). The measurement of suicidality, aggressivity and impulsivity. *Clinical Neuropharmacology*, **9**, 380-382.

Plutchik, R., & van Praag, H.M. 1990). A self-report measure of violence risk, II. *Comprehensive Psychiatry*, **31**, 450-456.

Plutchik, R., van Praag, H., & Conte, H.R. (1985). Suicide and violence risk in psychiatric patients. *Biological Psychiatry*, **20**, 761-763.

Plutchik, R., van Praag, H.M., Conte, H.R., & Picard, S.(1989). Correlates of suicide and violence risk I: the Suicide Risk measure. *Comprehensive Psychiatry*, **30**, 296-302.

Potegal, M., Yoburn, B., & Glusman, M. (1983). Disinhibition of muricide and irritability by intraseptal muscimol. *Pharmacology, Biochemistry and Behavior*, **19**, 663-669.

Pucilowski, O., Kostowski, W., Bidzinski, A., & Hauptmann, M. (1982). Effect of 6-hydroxydopamine-induced lesions of A10 dopaminergic neurons on aggressive behavior in rats. *Pharmacology, Biochemistry and Behavior*, **16**, 547-551.

Ramamoorthy, S., Bauman, A., Moore, K., Han, H., Yang-Feng, T., Chang, A., Ganapathy, V., & Blakely, R.D. (1993). Antidepressant- and cocaine-sensitive human serotonin transporter: molecular cloning, expression, and chromosomal localization. *Proceedings of the National Academy of Sciences of the USA*, **90**, 2542-2546.

Reis, D.J. (1972). The relationship between brain norepinephrine and aggressive behavior. *Research Publication of the Association for Research of Nerve and Mental Disorders*, **50**, 266-297.

Reza, A., Mercy, J.A., & Krug, E. (2001). Epidemiology of violent deaths in the world. *Injury Prevention*, **7**, 104-111.

Rudissaar, R., Pruus, K., Skrebuhhova-Malmros, T., Allikmets, L., & Matto, V. (2000). Involvement of GABAergic neurotransmission in the neurobiology of the apomorphine-induced aggressive behavior paradigm, a model of psychotic behavior in rats. *Methods and Findings in Experimental and Clinical Pharmacology*, **22**, 637-640.

Russ, M.J., Lachman, H.M., Kashdan, T., Saito, T., & Bajmakovic-Kacila, S. (2000). Analysis of catechol-O-methyltransferase and 5-hydroxytryptamine transporter polymorphisms in patients at risk for suicide. *Psychiatry Research*, **93**, 73-78.

Ryan, J.M. (2000). Pharmacologic approach to aggression in neuropsychiatric disorders. *Seminars in Clinical Neuropsychiatry*, **5**, 238-249.

Rydin, E., Schalling, D., & Asberg, M. (1982). Rorschach ratings in depressed and suicidal patients with low levels of 5-hydroxyindoleacetic acid in cerebrospinal fluid. *Psychiatry Research*, **7**, 229-243.

Sandler, M., Ruthven, C.R., Goodwin, G.L., Field, H., & Matthews, R. (1978). Phenylethylamine overproduction in aggressive psychopaths. *Lancet*, **2**, 1269-1270.

Shaffer, D. (1988). The epidemiology of teen suicide: an examination of risk factors. *Journal of Clinical Psychiatry*, **49**, 36-41.

Tardiff, K., & Sweillam, A. (1980). Factors related to increased risk of assaultive behavior in suicidal patients. *Acta Psychiatrica Scandinavica*, **62**, 63-68.

Turker, T., Sodmann, R., Goebel, U., Jatzke, S., Knapp, M., Lesch, K.P., Schuster, R., Schultz, H., Weiler, G., & Strober, G. (1998). High ethanol tolerance in young adults is associated with the low-activity variant of the promoter of the human serotonin transporter gene. *Neuroscience Letters*, **248**, 147-150.

Valzelli, L. (1981). *Psychobiology of aggression and violence*. New York: Raven Press.

van Praag, H.M. (1983). CSF 5-HIAA and suicide in non-depressed schizophrenics. *Lancet*, **2**, 977-978.

van Praag, H.M. (1991). The present and the future of serotonergic drugs. *Human Psychopharmacology*, **6**(Suppl), S13-S19.

van Praag, H.M., Lemus, C., & Kahn, R. (1986). The pitfalls of serotonin precursors as challengers in hormonal probes of central serotonin activity. *Psychopharmacological Bulletin*, **22**, 565-570.

Verkes, R.J., Pijl, H., Meinders, A.E., & van Kempen, G.M.J. (1996). Borderline personality, impulsiveness and platelet monoamine measures in bulimia nervosa and recurrent suicidal behavior. *Biological Psychiatry*, **40**, 173-180.

Weissman, M., Fox, K., & Klerman, G. (1973). Hostility and depression associated with suicide attempts. *American Journal of Psychology*, **130**, 450-455.

Woodman, D., & Hinton, J. (1978). Catecholamine balance during stress anticipation: an abnormality in maximum security hospital patients. *Journal of Psychosomatic Research*, **22**, 477-483.

Zalsman, G., Frisch, A., Bromberg, M., Gelernter, J., Michaelovsky, E., Campino, A., Erlich, Z., Tyano, S., Apter, A., & Weizman, A. (2001). Family-based association study of serotonin transporter promoter in suicidal adolescents: Possible role in violence traits. *American Journal of Medical Genetics*, **105**, 239-245.

Zalsman, G., Frisch, A., King, R.A., Pauls, D.L., Grice, D.E., Gelernter, J., Alsobrook, J., Michaelovsky, E., Apter, A., Tyano, S., Weizman, A., & Leckman J.F. (2001). Case-control and family-based association studies of Tryptophan Hydroxylase A218C polymorphism and suicidality in adolescents. *American Journal of Medical Genetics*, **105**, 451-457.

Name Index

Subject Index

www.ingramcontent.com/pod-product-compliance
Ingram Content Group UK Ltd.
Pitfield, Milton Keynes, MK11 3LW, UK
UKHW020135160926
475607UK00003B/64

9780792376743